MW00650605

USAF F-15 EAGLES

USAF F-15 EAGLES
Units, Colors & Markings

DON LOGAN

Schiffer Military History
Atglen, PA

ACKNOWLEDGEMENTS

I would like to thank Brian C. Rogers and Tom Kaminski for their help in the research for this book. Thanks to Roger Johansen for his work drawing the line art. Craig Brown, John Cook and Kevin Patrick supplied the Unit patches used in this book.

I would also like to thank the following individuals who provided photographs included in this book: Peter Becker, Tom Brewer, David F. Brown, Tony Cassanova, George Cockle, John Cook, Robert F. Dorr, Neil Dunridge, Gary Emery, Michael France, Alec Fushi, Jerry Geer, Jim Geer, Jim Goodall, Norris Graser, Bob Greby, Paul Hambleton, Paul Hart, Phillip Huston, Marty Isham, R.E.F. Jones, Dennis R. Jenkins, Tom Kaminski, Craig Kaston, Ben Knowles, Ray L. Leader, Bob Leavitt, Nate Leong, Kevin Patrick, William R. Peake, Doug Remington, Brian C. Rogers, Barry Roop, Mick Roth, Jim Rotramel, Tony Sacketos, Joe Sadler, Douglas Slowiak/Vortex Photo Graphics, Keith Snyder, Don Spering/AIR, Keith Svendsen, Jeff Wilson, Pete Wilson, Scott Wilson, Alan Van Winkle, and McDonnell Douglas/Boeing.

THE AUTHOR

After graduating from California State University-Northridge with a BA degree in History, Don Logan joined the USAF in August 1969. He flew as an F-4E Weapon Systems Officer (WSO), stationed at Korat RTAFB in Thailand, flying 133 combat missions over North Vietnam, South Vietnam, and Laos before being shot down over North Vietnam on July 5, 1972. He spent nine months as a POW in Hanoi, North Vietnam. As a result of missions flown in Southeast Asia, he received the Distinguished Flying Cross, the Air Medal with twelve oak leaf clusters, and the Purple Heart. After his return to the U.S., he was assigned to Nellis AFB where he flew as a rightseater in the F-111A. He left the Air Force at the end of February 1977.

In March of 1977 Don went to work for North American Aircraft Division of Rockwell International, in Los Angeles, as a Flight Manual writer on the B-1A program. He was later made Editor of the Flight Manuals for B-1A #3 and B-1A #4. Following the cancellation of the B-1A production, he went to work for Northrop Aircraft as a fire control and ECM systems maintenance manual writer on the F-5 program.

In October of 1978 he started his employment at Boeing in Wichita, Kansas as a Flight Manual/Weapon Delivery manual writer on the B-52 program. He is presently the editor for Boeing's B-52 Flight and Weapon Delivery manuals.

Don Logan is also the author of *Rockwell B-1B: SAC's Last Bomber; The 388th Tactical Fighter Wing At Korat Royal Thai Air Force Base 1972; Northrop's T-38 Talon; Northrop's YF-17 Cobra; Republic's A-10 Thunderbolt II; The Boeing C-135 Series – Stratotanker, Stratolifter, and other Variants; General Dynamics F-111 Aardvark; and ACC Bomber Triad: The B-52s, B-1s and B-2s of Air Combat Command.* (All available from Schiffer Publishing Ltd.)

Book Design by Robert Biondi.

Printed in China.
ISBN: 0-7643-1060-7

We are always looking for people to write books on new and related subjects. If you have an idea for a book, please contact us at the address below.

Published by Schiffer Publishing Ltd.
4880 Lower Valley Road
Atglen, PA 19310 USA
Phone: (610) 593-1777
FAX: (610) 593-2002
E-mail: Schifferbk@aol.com.
Visit our web site at: www.schifferbooks.com
Please write for a free catalog.
This book may be purchased from the publisher.
Please include $3.95 postage.
Try your bookstore first.

In Europe, Schiffer books are distributed by:
Bushwood Books
6 Marksbury Ave.
Kew Gardens
Surrey TW9 4JF
England
Phone: 44 (0)208 392-8585
FAX: 44 (0)208 392-9876
E-mail: Bushwd@aol.com.
Free postage in the UK. Europe: air mail at cost.
Try your bookstore first.

Contents

Introduction

McDonnell Douglas F-15 Eagle has been in operational US Air Force service since the 1st Tactical Fighter Wing received its F-15 aircraft at the beginning of 1976. Since then the US has needed to use it in two major conflicts, Operation Desert Storm and Operation Allied Force. In Desert Storm the USAF F-15s claimed 35 victories, 32 airplanes and 3 helicopters, without losing a single F-15 to another aircraft. In Operation Allied Force F-15Cs shot down four Serbian MiG-29s, four of the five air-to-air kills credited to U.S. forces during the conflict. Adding the victories claimed by the Saudi's in Desert Storm and Israeli's in other conflicts, the F-15 has scored over 100 air-to-air victories without a single loss to enemy aircraft.

During the F-15's USAF service, major changes in the operational structure of the US Air Force have occurred. The Aerospace Defense Command (ADC) was inactivated. With the disbanding of ADC the responsibility for continental air defense was taken over by Air Defense Tactical Air Command (ADTAC) and Air National Guard Units. The F-15s served a short career in four active Air Force ADTAC units before the responsibility was shifted to the Air National Guard. In 1991, Tactical Air Command (TAC), was replaced by the new Air Combat Command (ACC). TAC's F-15 units transferred to the new ACC. The F-15 has been assigned to Air National Guard units in seven states, six continue to operate Eagles.

The F-15E Strike Eagle has replaced the F-111 as the USAF's strike fighter. During Operation Allied Force F-15Es were the USAF's primary strike bomber. In addition to having the same air-to-air capability as the F-15C/Ds, the Strike Eagle can deliver all types of gravity non-nuclear weapons in the USAF inventory. The F-15E's LANTIRN system gives it a high speed-low altitude, terrain-following capability and also the ability to designate targets for laser guided weapon delivery.

(USAF via Marty Isham)

CHAPTER ONE

Program History

INITIAL DEVELOPMENT

The McDonnell Douglas F-15 Eagle has its origin back in the mid-1960s, when the U.S. aircraft industry was invited to study US Air Force requirements for an advanced tactical fighter that would replace the F-4 Phantom as the primary fighter aircraft in service with the USAF. Such an aircraft needed to be capable of gaining air superiority over any projected threats in the post-1975 period. Though primarily designed for air-to-air combat, the aircraft was also required to perform a secondary air-to-ground mission.

The Phantom II had originally been designed back in the 1950s to Navy requirements for a two-seat multi-role fighter, intended to destroy enemy aircraft at beyond-visual-range (BVR), using a powerful fire control radar to detect threats and to direct Sparrow semi-active radar guided missiles against them. No gun was provided, since in the late 1950s it was believed that the internal gun

was obsolete and not required in the missile age. Throughout much of the Vietnam War, the primary fighter in service with the USAF was the McDonnell F-4 Phantom II, a large, two-seat aircraft. The North Vietnamese Air Force was equipped with MiG-17s and MiG-21s, small, simple aircraft designed for close-in dog fighting. In 1965-68, the kill-ratio in air battles against the North Vietnamese Air Force was only 1.5 to one, much poorer results that those obtained in Korea by the F-86 Sabre against the MiG-15. A number of reasons accounted for this lower kill ratio. The North Vietnamese Air Force polished their tactics and engaged the U.S. forces only when it was in their favor. The U.S. forces had restrictive rules of engagement over North Vietnam, which required a close-in positive identification of the enemy before missiles could be fired. This negated the advantage of the Phantom II's radar and long-range Sparrow missiles. In a close-in fight against MiG-17s and MiG-21s, the Phantom II was less maneuverable and was at a relative disad-

The first F-15, 71-0280 (F1), is seen on June 26, 1972, at the rollout ceremony at McDonnell Douglas, St Louis, Missouri. (USAF) Place between title "INITIAL DEVELOPMENT" and text.

vantage in these types of encounters. The last three years of the war, the Phantom II became the USAF's primary air-to-ground fighter, taking the mission over from the F-105's. Most F-4 aircrews flew a vast majority of missions performing air-to-ground weapon delivery and were out of practice in the air-to-air mode. As a result, they were not proficient in exploiting the strengths of their own aircraft against the weaknesses of the enemy's planes.

The experience over North Vietnam caused the U.S. Air Force to conclude they needed to focus on the requirement for close-in air-to-air fighting in the design of their future fighter aircraft, not to rely just on superior radar and long-range missiles to ensure victory. At first the Air Force was rather uncertain of just what kind of aircraft they wanted to replace the Phantom II, and relied on the industry to tell them what the design should be rather than issuing any specific requirements.

On October 6, 1965, the Air Force issued Qualitative Operational Requirement (QOR) 65-14F, defining what later came to be known as the F-X (Fighter-Experimental) project. A Request For Proposal (RFP) was issued to the industry on December 8, 1965. The Air Force initially pictured the F-X as being a close-support multi-role aircraft powered by a pair of advanced turbofan engines and equipped with variable-geometry wings. Boeing, Lockheed, North American, Grumman, and McDonnell all wanted a piece of the action and began to work on initial concept studies.

After reviewing the initial concept studies, in March 1966 the USAF issued Concept Formulation Study (CFS) contracts to three of the five manufacturers – Boeing, Lockheed, and North American. Although Grumman and McDonnell had not been awarded any Air Force contracts, they nevertheless continued to fund their own studies on the same requirements.

None of the submitted designs were considered any further by the Air Force, mainly because the aerodynamic configurations and the bypass ration of the turbofans were considered inadequate. No prototypes were ordered and work on the F-X proceeded at a slow pace from mid-1966 too late 1967. On April 28, 1967, McDonnell merged with the Douglas Aircraft Corporation, becoming McDonnell Douglas, with all work on the company's F-X proposal continuing at their St. Louis facility.

On December 30, 1968 preliminary development contracts were awarded to Fairchild-Republic, McDonnell Douglas, and North American Rockwell. The North American Rockwell and Fairchild-Republic proposals both had single tail fins. The Fairchild-Republic proposal had its engines hanging out from the fuselage underneath a blended lift surface. The McDonnell Douglas proposal was a large, single-seat aircraft with twin vertical fins and a pair of turbofan engines. The FX was now known as the F-15.

On December 23, 1969, the McDonnell Douglas proposal was named the winner of the contest, and the company was authorized to proceed with the design and development phase, to build and test twenty Full Scale Development (FSD) aircraft, and to manufacture 107 single-seat F-15s and two-seat TF-15s.

Very early in the program, the McDonnell Douglas team had rejected the idea of using a variable-geometry wing as being too complex, too heavy, and too expensive. The team selected instead

The Day-Glo orange flight test markings applied to the standard air superiority blue underside of 72-0113 are visible in this photo. In addition to the three external fuel tanks, the FAST (Fuel and Sensor Tactical) conformal fuel tanks are visible on the side of the fuselage at the root of each wing. (USAF via Dennis R. Jenkins)

a large-area, fixed-geometry wing with a 45-degree sweep at the leading edge. The use of advanced avionics and electronics made it possible to use the single-seat configuration favored by the Air Force. The engines were to be a pair of Pratt & Whitney afterburning turbofans fed by lateral intakes. The air-to air armament was quite similar to that of the F-4 Phantom II, consisting of four AIM-7 Sparrow semi-active radar homing missiles mounted on the lower corners of the fuselage and four AIM-9 Sidewinder infrared-homing air-to-air missiles carried on wing stations. Like the F-4E, a 20-mm M61A1 cannon was installed, but in the right wing leading edge rather than the nose. Provision was incorporated in the F-15 to carry three 610-gallon drop tanks or up to 9000 pounds of air-to-ground stores, although, like the original F-4 design, the air-to-ground role was only secondary.

The F-15 was ordered with no prototype as such and no competitive fly-off against other manufacturer's aircraft. This raised quite a bit of controversy, with the press fearing another program with considerable cost overruns. As a result, and in response to criticism from Congress over cost overruns and lengthy delays that had occurred in both the C-5A Galaxy and F-111 programs, the USAF introduced a set of required demonstration milestones which the contractor had to meet before additional funding would be released. For the F-15 project, the milestones began with the preliminary design review, which was to be held by September 1970, and ended with a requirement that the first aircraft be delivered for test to the Air Force in November 1974.

(Boeing)

Total F-15 Eagles produced:

Development test aircraft

	F-15A	10
	TF-15A	2

Production USAF Aircraft

	F-15A	353
	F-15B	56
	F-15C	408
	F-15D	59
USAF Strike Eagles	F-15E	226

Total USAF F-15 Produced   1114

PROTOTYPE AND FLIGHT TEST F-15s

The F-15A was the first production version of the Eagle. There was no XF-15, since the aircraft had been ordered "off the drawing board." Including deliveries to non-USAF customers, a total of 384 single-seat F-15As were built, including 18 Full-Scale Development (FSD) aircraft.

The initial F-15 contract called for 20 FSD aircraft—a preliminary batch Category I aircraft made up of 10 single-seat F-15A (71-0280/0289) and two TF-15A two-seat (71-0290 and 71-0291) later designated F-15Bs. Category I flight tests are carried out with the manufacturer's test pilots and included all the basic tests required to demonstrate the airframe, engines, and all associated subsystems met the contractors specifications.

Eight Category II FSD aircraft, all of them in F-15A single-seat form (72-0113/0120) were also purchased. Category II testing involves flight testing by a USAF joint test force consisting of pilots from Air Force Systems Command and the Tactical Air Command. During this phase the Air Force pilots fly the aircraft and check the equipment to determine how well it meets combat requirements.

Category III testing is the follow-on operational test and evaluation program carried out by the primary using command (in this case Tactical Air Command) in the field. This phase of testing is also used to develop tactics and techniques, and determine operational restrictions. Also during Category III testing, logistics support requirements are established and personnel proficiency ratings are determined.

The first F-15A, serial number 71-0280, was rolled out in a ceremony at St. Louis on June 26, 1972. It was dismantled, loaded aboard a C-5A, and flown to Edwards AFB, California. It made its first flight July 27, 1972, with company test pilot Irving Burrows at the controls.

The first TF-15A (71-0290) was slotted between the seventh and eighth single seat aircraft. TF-15A 71-0291 was given a red-white-and-blue Bicentennial color scheme in 1976 and undertook a long round-the-world sales tour. In later years, 71-0291 served as the F-15E Strike Eagle demonstrator and was used to test the Eagle's proposed Fuel and Sensor Tactical (FAST) conformal fuel tanks.

Some of the 12 Category I test aircraft (10 single-seat and 2 two-seat aircraft) were later delivered to the Air Force. By October 29, 1973, 11 of the 12 Category I Eagles had entered flight testing. A maximum speed of Mach 2.3 and an altitude of 60,000 feet had been reached. Very few problems were encountered during flight testing. However, early in the test program, a problem was encountered – airframe buffeting at certain altitudes. To solve the problem, approximately four square feet of wing area diagonally from the wing tip was removed, giving the Eagle its characteristic raked wingtips. A flutter problem discovered during wind tunnel testing required a "dog tooth" be placed in the leading edge of the horizontal stabilator. The "dog tooth" change generated vortices and increased stabilator effectiveness, while curing flutter problems and eliminating buffet. The dorsal speed brake caused excessive buffeting when it was in the fully-open position. This was corrected by increasing the area of the speed brake from 20 to 31 square feet, allowing the required drag to be achieved without extending it as far into the airstream.

A majority of the test aircraft flew with day-glo orange markings on their wingtips, vertical and horizontal stabilizers, and engine inlets. This extra paint was designed to make them more visible during test flights. After the test program was completed, several of the early test aircraft received new gloss white paint jobs with a contrasting bright orange or royal blue trim color. The fourth F-15A (71-0283) received bright orange (but not day-glo) trim, with 71-0285, 71-0289, and 71-0290 using royal blue trim.

The first F-15B (71-0290) was later painted in a colorful red, white, and blue paint scheme for its use with NASA as the F-15S/MTD. The second F-15B (71-0291) was also painted red, white, and blue, duplicating a proposed paint scheme for the USAF Thunderbirds Air Demonstration Squadron should they ever fly F-15s. The aircraft was painted in late 1975 and 1976 to celebrate the US Bicentennial and later went on a company sponsored world-tour with a slightly modified version of the scheme. When used as the Strike Eagle demonstrator, 71-0291 was originally painted in Compass Ghost, then was repainted in the European One scheme of dark gray (FS36081) and two dark greens (FS34092 and FS34102).

F-15 TEST FORCE AIRCRAFT

Serial	#	First Flight	Function
71-0280	F1	July 27, 1972	Open the flight envelope, explore handling qualities, check out external stores carriage.
71-0281	F2	November 26, 1972	Tests of F100 engine.
71-0282	F3	November 4, 1972	Avionics development, calibrated air speed tests. First to be equipped with the APG-63 radar.
71-0283	F4	January 13, 1973	Flying structural test airframe.

(USAF)

71-0284	F5	March 7, 1973	First F-15 to be equipped with the M61A1 cannon. Internal gun, external fuel jettison and armament tests.	71-0291	T2	October 18, 1973	Last category I test ship, used by McDonnell Aircraft as a factory demonstrator for a world tour. Used as the Strike Eagle demonstrator.
71-0285	F6	May 23, 1973	Second avionics test aircraft. Avionics tests, flight control evaluation, missile fire control. (Nicknamed KILLER)	72-0113	F11		Operational Tests, FAST conformal fuel tanks.
				72-0114	F12		Operational Tests, later sold to Israel in 1992.
71-0286	F7	June 14, 1973	Armament and external fuel stores tests.	72-0115	F13		Operational Tests.
71-0287	F8	August 25, 1973	Spin recovery, high angle of attack, and fuel system tests.	72-0116	F14		Climatic tests, nicknamed "Homer", later sold to Israel.
				72-0117	F15		Operational Tests, later sold to Israel.
71-0288	F9	October 20, 1973	Integrated aircraft/engine performance tests.	72-0118	F16		Operational Test and demonstrations.
71-0289	F10	January 16, 1974	Tactical electronic warfare system, radar and avionics evaluation.	72-0119	F17		Streak Eagle – Time to Climb Records, retired to USAF Museum at Wright Patterson AFB, Ohio.
71-0290	T1	July 7, 1973	Initially designated as a TF-15A, it was the first two-seat F-15 manufactured.	72-0120	F18		Sold to Israel.

71-0280 F1

Right: The tail of the first F-15, 71-0280 carries the F-15 logo. (USAF)

Below: Seen in October 1974 carrying inert MK 82 500 pound bombs, the effect of the California desert sun on the Day-Glo paint is visible on the tail of 71-0280 (F1). (Dennis R. Jenkins)

71-0281 F2

71-0281 (F2) was loaned to NASA's Dryden Flight Research Center (DFRC). Initially only partial NASA markings, the white tail with the gold stripe and the NASA "meatball" on the nose, were applied. (Dennis R. Jenkins)

71-0282 F3

Above: 71-0282 carried the U.S. Bicentennial emblem on the outer surfaces of both vertical stabilizers. (Dennis R. Jenkins)

Right: 71-0282 (F3), the first F-15 to be equipped with the AN/APG-63 radar, seen here on February 2, 1976 was initially used for avionics development tests. (Dennis R. Jenkins)

Below: 71-0282 wore a sharkmouth on the nose, an eagle on the tail, and the ACES (Advanced Environmental Control System) on the left intake during 1977, while testing the new ACES. (Dennis R. Jenkins)

71-0283 F4

Above: 71-0283 (F4) was the first F-15 displayed to the general public. It is seen on the Edwards AFB flight line at the Edwards AFB Air Show in May 1973. (Don Logan)

Left: 71-0283 is seen on December 13, 1986 in the white and Day-Glo markings of the 6510th Test Wing. (Kevin Patrick)

Below: 71-0283 is seen here in formation with the number 5 F/A-18. (Boeing)

71-0284 F5

71-0284 (F5), the first fitted with the M61A1 20mm cannon, was used for armament and external fuel tank jettison tests. It had a relatively short career, seen here on February 3, 1976 minus its rudders and speed brake. (Dennis R. Jenkins)

71-0285 F6

Below: 71-0285 (F6) is seen here taxiing out for a test mission on February 3, 1976. (Dennis R. Jenkins)

71-0285 seen in July 1982 at Offutt AFB, Nebraska in blue and white test markings. (George Cockle)

71-0286 F7

Like 71-0284 (F5), 71-0286 (F7) seen here at Edwards AFB on January 16, 1976, was used for armament and fuel tank jettison tests. (Dennis R. Jenkins)

71-0286 is seen taxiing out for a test mission on February 17, 1976, (Dennis R. Jenkins)

71-0287 F8

71-0287 (F8), the first Day-Glo and white F-15 seen here on February 27, 1976 was used in spin recovery, high angle of attack and fuel system testing. (Dennis R. Jenkins)

Right: 71-0287 is seen dumping fuel as part of fuel system testing. (USAF)

Below: 71-0287 is seen being prepared for a test flight. AIM-7 Sparrows are visible on the fuselage stations. (Tom Brewer Collection)

71-0288 F9

71-0288 (F9) seen on February 27, 1976 was used for integrated aircraft and engine performance tests. (Dennis R. Jenkins)

71-0289 F10

71-0289 (F10) was used for tactical electronic warfare system, radar and avionics evaluation. (Don Logan Collection)

71-0289 is seen in overall air superiority blue after removal of the high visibility Day-Glo flight test markings. (Don Logan Collection)

71-0289 seen on September 14, 1982 at Forbes Field, Kansas, in blue and white markings. (Don Logan)

71-0290 T1

71-0290 (T1) seen here on February 7, 1974 taxiing for a test mission was the first TF-15A (later redesignated as a F-15B) (Dennis R. Jenkins)

71-0290 had been repainted without Day-Glo when seen here taxiing for a test mission at Edwards AFB on March 26, 1976. (Dennis R. Jenkins)

71-0290 seen on May 26, 1983 at Luke AFB, Arizona, in blue and white markings. (Kevin Patrick)

71-0291 T2

71-0291 (T2), the last Category 1 test aircraft, is seen carrying the insignia of the Air Force Flight Test Center (AFFTC). (Don Logan Collection)

71-0291 is seen here at Edwards AFB with a silver radome carrying the long flight test alpha/beta probe. (Dennis R. Jenkins)

For four days in April 1976, the second TF-15A (T2), 71-0291, was painted in Armée de l'Air (French Air Force) markings for demonstration flights. The French decided not to purchase the F-15. (Dennis R. Jenkins)

Above: During late 1975 McDonnell Douglas repainted 71-0291 in a colorful red white and blue scheme with the U.S. Bicentennial emblem added during 1976. While in this paint scheme, the aircraft visited numerous foreign countries flying demonstration flights for foreign air forces. (Don Logan Collection)

Right: Each time the aircraft visited a new country, the country's flag was added to the left side of the fuselage below the cockpit. (Dennis R. Jenkins)

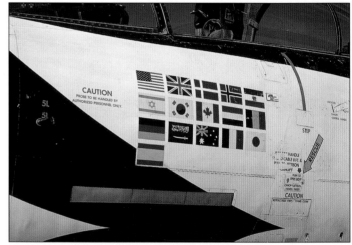

Below: 71-0291 flew missions both with and without the long radome mounted flight test probe seen here. (Don Logan Collection)

71-0291 continued flight test duties. It is seen in the standard Compass Ghost paint scheme carrying AIM-120 AMRAAM missiles. (Boeing)

71-0291 was repainted in a modified "Europe One" paint scheme while used as the Strike Eagle demonstrator. (Boeing)

71-0291, as the Strike Eagle demonstrator is loaded with MK 20 Rockeye Cluster Bombs. (Don Logan Collection)

72-0113 F11

72-0013 (F11), the first 1972 F-15, is seen here at Edwards AFB carrying the insignia of the Air Force Flight Test Center (AFFTC). (Tom Brewer Collection)

72-0113 is seen with 72-0114 (F12) on July 2, 1974 at Andrews AFB, Maryland in the air superiority blue paint. It carries both the Air Force Systems Command badge and the TAC Badge. Also of note is the light gray serial number on the tail of 72-0113 compared to the black serial of 72-0114 in the background. (Tom Brewer Collection)

72-0114 F12

72-0114 (F12) is seen on July 2, 1974 at Andrews AFB, Maryland in the air superiority blue paint. It carries both the Air Force Systems Command badge and the TAC Badge. (Tom Brewer Collection)

72-0116 F14

72-0016 (F14) was nicknamed "Homer". It was used for climatic testing, spending considerable time in the environmental hangar at Eglin AFB. It was later delivered to Israel as part of the "Peace Fox I" program. (Don Logan)

STREAK EAGLE

During 1974 and 1975, McDonnell modified F-15A serial number 72-0119 in order to set world time-to-climb records. The project was given the name Operation Streak Eagle. In an effort to save weight, all non-mission critical systems were deleted, including flaps, speed brake, armament, radar, and fire control system. The aircraft was finished in bare metal, with no paint. It weighed 1800 pounds less than the stock F-15A. The record attempts were carried out during the winter at Grand Forks AFB, North Dakota to take advantage of the cold temperatures. During the record attempts, only enough fuel was carried to make the specific flight and return to base. The aircraft broke eight existing time-to-climb records previously held by the F-4B and the MiG-25 (see chart below).

(USAF)

WORLD TIME-TO-CLIMB RECORDS

ALTITUDE (M)	TIME (SEC)	PILOT	DATE	PREVIOUS RECORD (SEC)
30,000	207.80	MAJ R. SMITH	1 FEB 75	243.86 (MIG-25)
25,000	161.02	MAJ D.W. PETERSON	26 JAN 75	192.60 (MIG-25)
20,000	122.94	MAJ R. SMITH	19 JAN 75	169.80 (MIG-25)
15,000	77.04	MAJ D.W. PETERSON	16 JAN 75	114.50 (F-4)
12,000	59.38	MAJ W.R. MACFARLANE	16 JAN 75	77.14 (F-4)
9,000	48.86	MAJ W.R. MACFARLANE	16 JAN 75	61.68 (F-4)
6,000	39.33	MAJ W.R. MACFARLANE	16 JAN 75	48.79 (F-4)
3,000	27.57	MAJ R. SMITH	16 JAN 75	34.52 (F-4)

Most of these records were later broken by the Soviet P-42, which was a prototype for the Sukhoi Su-27 Interceptor. The Streak Eagle has been painted and is now on outdoor display at the USAF Museum at Wright Patterson AFB at Dayton, Ohio.

72-0119 F17 STREAK EAGLE

72-0119 was not used it the flight test program. It was used in Project Streak Eagle to set world time-to-climb records during 1975. To save weight, the aircraft was not painted. (USAF)

72-0119 is seen here at Grand Forks AFB, North Dakota being prepared for a record setting mission. The missions were flown during the winter and from Grand Forks to take advantage of the extra thrust generated in the colder, denser air. (Douglas Slowiak/Vortex Photo Graphics)

72-0119 is seen here at Edwards AFB, being prepared for Project Streak Eagle. (Tom Brewer)

F-15 ASAT

In the late 1970s, even before the advent of the Strategic Defense Initiative (SDI) also called the "Star Wars" program, an anti-satellite (ASAT) capability evolved for the F-15. The goal of ASAT weapons was to destroy enemy military satellites, particularly low-orbiting reconnaissance, ELINT (Electronic Intelligence), and ocean surveillance satellites. The Soviets had their own anti-satellite program in which a killer satellite would rendezvous with the target satellite and explode. The American equivalent involved the arming of an F-15 Eagle with a missile which would be launched against an orbiting satellite from a zoom climb at an altitude of 80,000 feet.

In 1979, a contract was issued to Vought for an air-launched low Earth-orbit anti-satellite vehicle. This missile, the Vought ASM-135A, was a two-stage rocket. The first stage was a modified AGM-69 SRAM-A and the second stage an Altair III rocket. The ASM-135A weighed about 2700 pounds at launch and was 18 feet long. The payload consisted of a miniature kinetic kill vehicle using an infrared seeker to home in on the target satellite, destroying it by impact. F-15A 76-0086 was modified for trials with the ASM-135A. It was carried on the F-15 centerline station. The first actual launch of an ASM-135A from an F-15 took place in January 1984. The

missile was aimed at a predetermined point in space. Three additional launches were made against celestial infrared sources. The first and only launch against an actual target satellite took place on September 13, 1985. On that date F-15A 77-0084 of the 6512th Test Squadron stationed at Edwards AFB took off from Vandenberg AFB and zoom-climbed up to 80,000 feet, launching the ASAT against the Solwind P78-1, a gamma ray spectroscopy satellite in orbit since February 1979. Both the first and second stages fired successfully, and the miniature kinetic kill vehicle separated and homed in on the satellite, successfully destroying it upon impact.

The test was a success in that it demonstrated that the basic concept was feasible. However, it enraged arms control advocates, who saw the test as a violation of a joint US/Soviet treaty limiting the development and testing of anti-satellite weapons. Initial plans had been made to modify twenty F-15As for the anti-satellite mission and to assign them to the 48th Fighter Interceptor Squadron (FIS) at Langley AFB in Virginia and the 318th FIS at McChord AFB in Washington. These squadrons had each received three or four F-15A/B airframes modified for ASAT operations. However, Congress was unwilling to permit any further testing of the system, and the ASAT program was officially terminated in 1988.

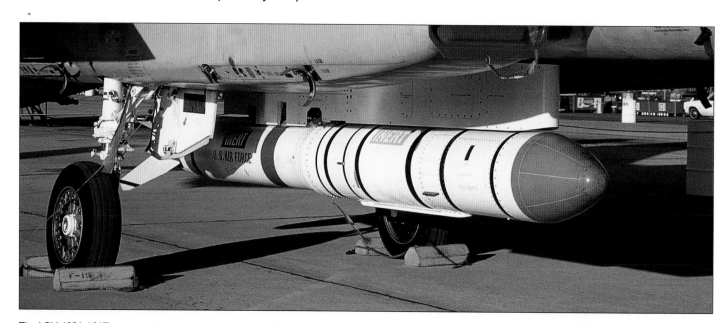

The AGM-135A ASAT was carried on the centerline of the F-15. It was a two stage missile using a AGM-69A SRAM missile as the first stage and the second stage from an Altair III rocket. (Dennis R. Jenkins)

"CELESTIAL ONE" 76-0084 was one of two F-15As used in the ASAT Test Program. The two aircraft carried different ASAT emblems on the tail. (Craig Kaston)

(Craig Kaston)

(Don Logan Collection)

76-0086 was the other ASAT program test aircraft. (Don Logan Collection)

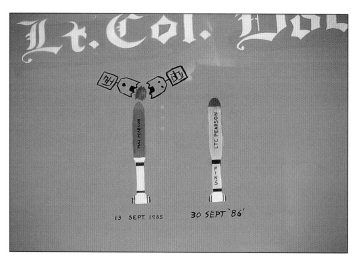

These launch markings were added to the left side of 76-0084 under the cockpit. They document the two "live" firings occurring in September 1985. (Craig Kaston)

F-15 WITH NASA

Two Category I F-15As, 71-0281 and 71-0287, were loaned to NASA in 1975 and 1976 and assigned to the NASA Dryden Flight Research Center at Edwards AFB. They have been used for several experimental programs.

71-0281 was acquired on December 17, 1975, and was used for the aerodynamic testing of the shuttle's thermal protection tiles by mounting them on the inner wing leading edge. The tiles attached to the right wing simulated those on the leading edge of the orbiter's wing, with the tiles on the left wing simulating those on the junction of the orbiter's wing and fuselage. The tiles attached to the F-15A's wing were ultimately subjected to almost 1.5 times the dynamic pressure, which the Shuttle experiences

71-0287 is seen on a NASA test mission in formation with a NASA F-104 and the Number 1 (72-1569) Northrop YF-17. (NASA via Dennis R. Jenkins)

during launch. 71-0281 was returned to the USAF on October 28, 1983 without ever being assigned a NASA number. It is now on display on a pylon with other single seat fighters at Langley AFB, Virginia.

71-0287 was acquired on January 5, 1976. It was assigned the NASA number of 835. It participated in a series of tests that involved the Digital Electronic Engine Control and other advanced engine features that were planned for the Pratt & Whitney 1128 turbofan, a derivative of the F100-PW-100, which ultimately led to the development of the F100-PW-220. Later, it participated in the testing of the NASA/USAF Highly Integrated Electronic Control (HIDEC) program which involved a flight control system that was capable of detecting in-flight failures and automatically reconfiguring the aircraft's control surfaces to compensate for them.

71-0290 was modified to the F-15S/MTD (STOL/Maneuvering Technology Demonstrator) configuration adding forward canards mounted on the inlets and two dimensional thrust vectoring exhaust nozzles on the engines. Beginning with its first flight in April of 1996, a further modified 71-0290 was used in the NF-15B ACTIVE (Advanced Control Technology for Integrated Vehicles program. The program modified the aircraft with special thrust vectoring F100-PW-229 engines.

Early in 1994 NASA began operating 74-0141 as an Aerodynamic Flight Facility with a black centerline pylon called the FTF II (Flight Test Fixture II). This pylon can be modified to carry various test payloads.

71-0287, wearing the NASA number 835, took part in the Highly Integrated Electronic Control (HIDEC) program. (NASA via Dennis R. Jenkins)

71-0281, seen here in NASA markings on February 21, 1980, was used to test Space Shuttle thermal protection tiles in 1980. It was never given a NASA number. (Dennis R. Jenkins)

71-0290 (T1) was modified into the F-15S/MTD (STOL/Maneuvering Technology Demonstrator) configuration adding forward canards on the inlet shoulders. It was later modified into the NF-15B ACTIVE (Advanced Control Technology for Integrated Vehicles) aircraft. (David F. Brown)

The forward canards on 71-0290 seen here were made from modified F/A-18 stabilators. The original small speed brake visible on the top of the fuselage was retained. (NASA via Dennis R. Jenkins)

The two-dimensional exhaust nozzles thrust vectoring/reversing nozzles are visible in this view of 71-0290 in the F-15S/MTD (STOL/Maneuvering Technology Demonstrator) configuration. (NASA via Dennis R. Jenkins)

The Flight Test Facility II (FTF II) is visible on the centerline of 74-0141, NASA 836 in this photo. The FTF II is capable of carrying a variety of experiments. (NASA via Dennis R. Jenkins)

CHAPTER TWO

F-15 Models

F-15A

Production of an initial batch of 30 F-15A/B fighters was announced in March 1973. The first Eagle delivered to an operational USAF unit was TF-15A 73-0108, a two seat model nicknamed TAC-1. TAC-1 was formally accepted on November 4, 1974 by the 555th Tactical Fighter Training Squadron, 58th Tactical Training Wing at Luke AFB, Arizona in a ceremony presided over by then-President Gerald Ford. The 58th TTW served as the Replacement Training Unit (RTU) for F-15 pilot training during the initial phases of the Eagle's entry into operational service. Ten Full System Development (FSD) and 384 production F-15As were built (353 of the 384 were assigned to the USAF).

The F-15 entered USAF service with a few problems. The pilots at Luke AFB found that the F-15 was unable to fly the planned number of sorties. The major problem was the Pratt & Whitney F-100-PW-100 engines. The Air Force had underestimated the number of engine power cycles per sortie and had not realized how often the Eagle's maneuvering capabilities resulted in frequent abrupt changes in throttle setting. These frequent abrupt changes caused unexpectedly high wear on key engine components, resulting in frequent failures of these components. However, the most serious engine problem was stagnation stalling.

There were frequent groundings and delays in engine deliveries while an attempt was made to fix these engine problems. By the end of 1979, the USAF was forced to accept engineless F-15 airframes and place them in storage awaiting sufficient numbers of engines. A massive effort by Pratt & Whitney helped to alleviate this problem, but still the F-15 suffered from an engine shortage for a long period of time.

Design modifications and improvements in materials, maintenance, and operating procedures overcame early problems with the reliability of the F100-PW-100 engine. The installation of a quartz window in the side of the afterburner assembly enabled a flame sensor to be installed to monitor the pilot flame of the afterburner. This helped to cure a problem of hard afterburner light-offs. Modi-

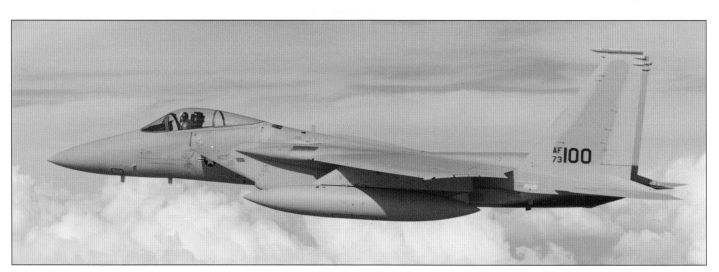

(Boeing via Marty Isham)

fications to the fuel control system helped to lower the frequency of stagnation stalls. In 1976 the F-15 fleet had suffered 11-12 stagnation stalls per 1000 flying hours. By the end of 1981, this rate was down to 1.5. However, the F100 continued to have a reputation of being a temperamental engine under certain conditions.

F-15As were assigned to ADTAC's Fighter-Interceptor Squadrons for air defense of the United States. The role of the Eagle in the air defense of the United States as part of ADTAC was a brief one. The 1st Air Force's Eagle interceptor squadrons were inactivated by the end of 1991. The 1st Air Force, along with its mission and many of its aircraft, was reassigned to the Air National Guard.

Under the Multi-Stage Improvement Program (MSIP), upgrades were progressively retrofitted into the F-15A. The F-15A models going through MSIP, though not fitted with the conformal fuel tanks, were externally indistinguishable from the F-15C. However, the very early F-15As (from Fiscal Years 1973, 1974, and 1975) were not upgraded under MSIP and were retired to the Aerospace maintenance and Regeneration Center (AMARC) at Davis Monthan AFB, Arizona. Some have been made available as gate guards or donated to museums, with still others given to Israel as payment for policy decisions made during the Gulf War.

TF-15A/F-15B TWO-SEAT TRAINER

The F-15B was the two-seat training version of the F-15A. It retained the basic airframe of the F-15A without extensive structural alterations and without any changes in the overall dimen-

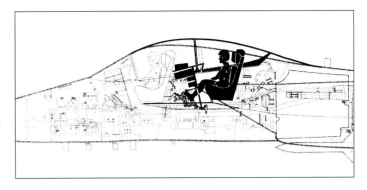

sions. Initially designated TF-15A, the F-15Bs differed from the F-15A primarily in having a second seat underneath a larger canopy. Although the two seat models were designed to be training aircraft, they are fully combat capable and have the same weapon system as the single seat aircraft. The only visible difference between the single seat and two seat F-15 is the "bumped-up" canopy. The canopy was enlarged in the aft area to allow the back seat occupant to have adequate head room and excellent visibility. Equipment installed in the aft cockpit to support the second crew member includes an ejection seat, flight controls, throttles, life support systems, instrument panel and consoles equipped with necessary flight and engine instruments and system controls. The B model did not have the ALQ-135 of the F-15A. Performance of the F-15B is essentially the same as the single seat aircraft. The F-15B weighs approximately 800 pounds more. A total of 2 FSD and 56 production F-15Bs were built.

71-0291 in McDonnell Douglas demonstrator markings is seen over St. Louis, Missouri. (Boeing)

Opposite: An F-15C goes vertical. (Boeing)

F-15C

The F-15C was the second single-seat version of the Eagle. The first F-15C (78-0468) flew its maiden flight on February 26, 1979. Delivery of the improved F-15C to the USAF began later that year. The 18th TFW at Kadena AB in Okinawa received F-15Cs, making the first Pacific unit to be assigned the Eagle. The first USAFE unit to receive the F-15C was the 32nd Tactical Fighter Squadron based at Soesterberg in the Netherlands. They replaced the unit's F-15A/Bs. A majority of the F-15C had the added capability of carrying Conformal Fuel Tanks (CFTs) attached to the side of the fuselage under the wings. A total of 408 F-15Cs were delivered to the USAF.

The F-15C had significant improvements to the avionics. The APG-63 or APG-70 radar of the F-15C is equipped with a Programmable Signal Processor (PSP), a special-purpose computer which controls the radar modes through software, allowing rapid switching of radar modes.

Most F-15Cs were delivered with Pratt & Whitney F100-PW-100 turbofans, and some were later re-engined with more reliable but slightly lower rated F100-PW-220 engines.

F-15D

The F-15D is the two-seat version of the F-15C. The F-15D is dimensionally identical to the F-15C and carries the same avionics suite with the same armament capabilities. Like the F-15B it lacked the ALQ-135 installed in the single seat Eagles It has essentially the same performance as well. The first F-15D (78-0561) took off on its maiden flight on June 19, 1979. A total of 59 F-15Ds were delivered to the USAF.

F-15E STRIKE EAGLE

In the late 1970s, McDonnell Douglas and Hughes Aircraft teamed in a privately-funded study of the feasibility of adapting the basic F-15 Eagle to the air-to-ground role. Originally conceived as a multi-role aircraft, by 1975 the F-15's air-to-ground role was set aside in favor of the air-to-air role. As part of this program, the McDonnell Douglas converted the second F-15B (71-0291) to the new Strike Eagle. The Strike Eagle first flew on July 8, 1980, and was initially equipped with a modified APG-63 radar, which used synthetic aperture radar techniques to do high-resolution ground mapping. The back seat was configured for a Weapon System Officer (WSO) operating the weapon delivery systems. The aircraft was equipped with the Conformal Fuel Tanks (CFT) first introduced on the F-15C/D, with six stub pylons on the lower corners and on the bottom of each of the CFTs for the carriage of bombs. The prototype tested a centerline gun pod and a Pave Tack laser designator pod (carried by some F-4Es and F-111Fs) on the left side of the forward air intake. The Pave Tack pod made the aircraft capable of delivering laser-guided weapons without the assistance of a separate designator aircraft.

The production version of the Strike Eagle was designated F-15E, with full scale development beginning in 1984. The first production F-15E (86-0183) made its first flight on December 11, 1986, with company test pilot Gary Jennings at the controls. The F-15E externally is similar to the two-seat F-15D. However, the F-15E is internally redesigned with a stronger structure to safely operate at weights as high as 84,000 pounds. The F-15E's structural life is 16,000 hours, more than twice that of earlier F-15s.

The F-15E was initially powered by 24,000 pounds static thrust afterburning Pratt & Whitney F100-PW-220 turbofans. The engine bays were adapted so that these engines could be replaced by more powerful turbofans in the 30,000 pound thrust class. Plans to deliver F-15Es with the twenty-percent more powerful F100-PW-229 engine beginning in August 1991 were delayed slightly. The -229 turbofans were installed in succeeding F-15Es, a majority of the F-15Es have -220E engines.

Following completion of operational test and evaluation at Edwards AFB and Seek Eagle weapons carriage and separation tests carried out at Eglin AFB, F-15Es were first delivered to the 405th TTW at Luke AFB for crew training. The first operational F-15E squadron was the 336th TFS, 4th TFW at Seymour-Johnson AFB in North Carolina, which received its first aircraft in early 1989. Still in production, so far a total of 226 F-15Es have been built for the USAF.

86-0183 was the first F-15E Strike Eagle. (Boeing) Place between title "F-15E STRIKE EAGLE" and text.

CHAPTER THREE

Test Units

AIR FORCE FLIGHT TEST CENTER (AFFTC)
EDWARDS AFB, CALIFORNIA

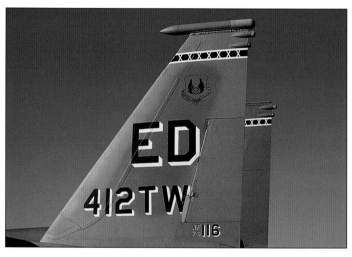

(Keith Snyder)

6510th TEST WING

The initial F-15 testing was accomplished by the F-15 Joint Test Force (JTF), part of the 6510th Test Wing, 6515th Test Squadron at Edwards AFB, California. Aircraft of the JTF did not have any unit tail markings. The 6515th flew F-15As and Bs and operated a few F-15C/Ds and F-15Es. Like other 6510th Test Wing aircraft, F-15s of the 6515th Test Squadron wore blue tail stripe with white Xs and no tail codes.

412th TEST WING

On October 1, 1992 the 6510th Test Wing was redesignated the 412th Test Wing, and the squadron redesignated the 415th Test Squadron, adding ED (EDwards AFB) tail codes. The 415th Test Squadron became the 415th Flight Test Squadron on March 1, 1994. On October 1, 1994, the 415th Flight Test Squadron was inactivated and the 445th Flight Test Squadron took over the F-15s.

Right: F-15A 76-0116 is seen here at Edwards AFB on October 2, 1998 marked as the 412th Test Wing Flagship. (Keith Snyder)

Below: F-15A 77-0139 is seen here at Edwards AFB wearing sliver F-22 test paint. (Keith Snyder)

F-15B 76-0140 is seen here at Edwards AFB in the white and red-orange paint scheme used by Test Pilot School and Flight Test Safety Chase aircraft. (Keith Snyder)

F-15D 79-0013 is seen on June 25, 1985 parked on the F-15 Test Facility ramp at Edwards AFB. (Craig Kaston)

This nose art is carried by F-15B 76-0134. (Craig Kaston)

This nose art is carried by F-15B 76-0132. (Craig Kaston)

F-15B 76-0130 is seen taxiing back from a test mission at Edwards AFB in the white and red-orange paint scheme on October 2, 1998. (Keith Snyder)

Above: F-15B 77-0166 is seen here at Robins AFB, Georgia in the Project Firefly markings. Project Firefly was the name assigned to the Integrated Flight-Fire Control (IFFC) system program. (Don Logan Collection)

Right: This view of 78-0468, the first F-15C, shows CFT (Conformal Fuel Tanks) fitted with the Strike Eagle bomb racks. An early LANTRIN pod is visible on the left forward station. (USAF via Robert F. Dorr)

Below: 77-0084, seen here on the ramp at Edwards AFB in May 1980, was used as a test bed for the AN/APG-63 Radar. (Alec Fushi)

71-0291, seen here in June 1990 in standard Compass Ghost paint scheme, carries "Peep Eagle" reconnaissance demonstrator emblem on the tail. The reconnaissance system was normally carried as a conformal centerline pod and had been removed when this photo was taken. (David F. Brown)

F-15D 82-0046 is seen taxiing back following a flight at Edwards AFB. (Keith Snyder)

F-15D 84-0046, seen here on October 19, 1997 the day after General Chuck Yeager used the aircraft to break the sound barrier to commemorate the 50th anniversary of his, and the world's first supersonic flight on October 18, 1974 in the Bell X-1. 84-0046 carried the name "Glamorous Glennis", the same name carried by the Bell X-1. (Craig Kaston)

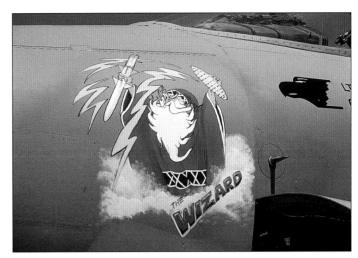

Right: 84-0046 carried nose art of the "Wizard". (Craig Kaston)

Below: This photo of F-15D 84-0046 taken on August 26, 1995, and the one below illustrate the two types of tail code/serial number makings carried on F-15s. (Craig Kaston)

F-15D 84-0046 seen here in July 1992, carries the low visibility tail markings (Ben Knowles)

F-15B 76-0140 is seen flying safety chase with F-22 number 2. (USAF)

F-15E 86-0190 carries a full load of 12 MK 82 500 pound bombs on its CTFs, four AIM-9 Sidewinders and a 600 gallon centerline drop tank. The LANTRIN pods are visible under the engine intakes. (Boeing)

Above: A full load of MK 20 Rockeye Cluster Bombs are visible in this view of 71-0291, the Strike Eagle demonstrator. (Boeing)

Left: 86-0183 was the first F-15E Strike Eagle. (Alec Fushi)

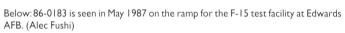

Below: 86-0183 is seen in May 1987 on the ramp for the F-15 test facility at Edwards AFB. (Alec Fushi)

87-0180 assigned to the 412th test wing is seen here at Edwards AFB on October 2, 1998. (Keith Snyder)

This F-15E, 91-0600, seen here on October 21, 1995, painted with a modified Compass Ghost paint scheme rather than the standard F-15E all gunship gray paint. (Craig Kaston)

AIR ARMAMENT CENTER (AAC)
EGLIN AFB, FLORIDA

The Air Armament Center (AAC) of Air Force Materiel Command, located at Eglin AFB, Florida has the mission of planning, directing, and conducting the test and evaluation of non-nuclear munitions, electronic combat, and navigation/guidance systems. AAC was formerly the Air Force Development Test Center (AFDTC), and prior to being AFDTC it was known as the Armament Development Test Center (ADTC) AAC is also responsible for all host and base support functions for Eglin AFB. The current flying unit for the Center is the 46th Test Wing. The 46th Test Wing replaced the 3246th Test Wing on October 1, 1992.

F-15C 84-0018 in 3246th Test Wing markings is seen launching a AIM-7 Sparrow missile. (USAF)

3246th TEST WING

The 3246th Test Wing and its 3247th Test Squadron operated the F-15 from its entry to USAF service. The 3246th Test Wing was replaced by the 46th Test Wing and the 3247th Test Squadron was redesignated as the 40th Test Squadron on October 1, 1992. Initially, the 3246th TW aircraft carried AD tail codes for Armament Division and a white tail stripe with red diamonds. In November 1989 the tail code was changed to ET for Eglin Test, with no change to the tail stripe.

This F-15D 80-0055, carrying a non-standard serial number presentation, is seen during a test launch of an AIM-120 AMRAAM. (Boeing)

F-15C 84-0011 in the markings of the 3246th Test Wing is seen here at Davis-Monthan AFB, Arizona on January 7, 1989. (Wally Van Winkle)

F-15C 84-0018 in 3246th Test Wing markings is seen here at Davis-Monathan on January 23, 1987. (Brian C. Rogers)

F-15E 86-0188 is seen over Eglin AFB, Florida Ranges in May 1989. The orange pylons on the outer wings and under the engines hold cameras to film weapon releases. (Boeing)

46th TEST WING

The 46th Test Wing conducts test and evaluation in support of AFDTC. The wing was established on February 10, 1975 as the 46th Aerospace Defense Wing and activated at Peterson AFB, Colorado on March 15, 1975. The 46th Flying Training Squadron, the wing's flying unit flew T-33 and T-37 aircraft. It was inactivated on October 1, 1983. The 46th was redesignated as the 46th Test Wing and activated at Eglin AFB on October 1, 1992, replacing the 3246th Test Wing. The 40th Flight Test Squadron is the 46th Test Wing's base flying squadron, which operates the F-15. The 39th Flight Test Squadron, also a part of the 46th Test Wing, operates the A-10 and F-16.

40th FLIGHT TEST SQUADRON

The 40th Test Squadron was redesignated the 40th Flight Test Squadron (FLTS) on March 15 1994. It operates F-15C/D and F-15Es from Eglin AFB. The 40th FLTS carries the ET tail code and the white tail stripe with red diamonds.

(Tony Cassanova)

F-15D 84-0045 is seen on May 15, 1998 marked as the 40th Flight Test Squadron Flagship. (Norris Graser)

F-15A 76-0101 is seen on October 22, 1996 at Eglin AFB, Florida with flagship type tail markings. (Peter Becker)

F-15A 76-0101 is seen in May 1998 at Cannon AFB, New Mexico. (Don Logan)

F-15A 76-0116 is seen in October 1994 at Eglin AFB, Florida in the markings of the 46th Test Wing. (Peter Becker)

F-15B 75-0084 is seen here at McConnell AFB in June 1996. (Jerry Geer)

F-15B 77-0161 is seen in May 1991. (David F. Brown)

The first F-15C, 78-0468, is seen in March 1993 in the markings of the 46th Test Wing. (Ben Knowles)

F-15D 84-0045 seen here in November 1990 assigned to the 3246th Test Wing carrying ET tail codes. (Ben Knowles)

F-15E 86-0185 in October 1994 in the markings of the 46th Test Wing. (Peter Becker)

46th TEST GROUP

The 46th Test Group manages the 586th Flight Test Squadron based at Holloman AFB, New Mexico operating HT tail coded (Holloman Test) YF-15A 71-0289, the tenth, and last Developmental Test F-15A.

F-15A number 10 71-0289 is seen in the markings of the 586th Flight Test Squadron, 46th Test Group with Holoman AFB HT tail codes. (Keith Snyder)

USAF AIR WARFARE CENTER

The center was originally activated as the USAF Tactical Air Warfare Center on November 1, 1963, at Eglin AFB, Florida, to conduct operational test and evaluation (OT&E) of tactical aircraft, air-to-air weapons, air-to-ground weapons, weapons delivery tactics, techniques, and procedures, and avionics. Major systems developed for operational use by Air Force forces included laser designators; various laser-guided bombs; the GBU-15 precision-guided glide bomb; and the Low Altitude Navigation and Targeting Infra-Red for Night (LANTIRN) pod systems used on F-15E and F-16C/D fighters. F-15 aircraft were operated by the 4443rd Test and Evaluation Group, 4485th Test Squadron. The aircraft wore OT (Operational Test) tail codes with a black and white checked tail stripe. TAWC was redesignated as the USAF Air Warfare Center (AWC) in October 1991. On December 1, 1991 the 4443rd Test and Evaluation Group was redesignated as the 79th Test and Evaluation Group, with the 4485th Test Squadron being redesignated as the 85th Test and Evaluation Squadron. The aircraft kept the OT tail codes and black and white checked tail stripe.

In October 1995, as part of Air Combat Command's restructuring, its total operational test and evaluation, and weapons and tactics training communities were merged into one center. The USAF AWC remained at Eglin AFB and was redesignated 53rd Wing, assigned as a component of the new Air Warfare Center, to be located at Nellis AFB. The roles and missions of the two bases (Eglin AFB and Nellis AFB) were combined into a single center to improve communication, efficiency and combat capability in the test and development process.

(Boeing)

4443rd TEST AND EVALUATION GROUP

As part of the TAWC, the 4443rd Test and Evaluation Group (TEG) operated F-15s until being redesignated the 79th Test and Evaluation Group on December 1, 1991.

4485th TEST SQUADRON

The 4485th Test Squadron operated F-15s as part of the 4443rd TEG until it was replaced by the 85th Test and Evaluation Squadron on December 1, 1991.

53rd WING

The 53rd Wing, located at Eglin AFB is the focal point for the combat air forces in electronic combat, armament and avionics, chemical defense, reconnaissance, command and control, and aircrew training devices. The Wing reports to the Air Warfare Center at Nellis AFB, Nevada. The 53rd Wing is also responsible for operational testing and evaluation of new equipment and systems proposed for use by these forces. In a realignment of operational test and evaluation functions, the 53rd Wing gained the 57th Test Group, located at Nellis AFB, Nevada, and all its component units, including the 422nd Test and Evaluation Squadron, on October 1, 1996.

53rd TEST AND EVALUATION GROUP

The 53rd Test and Evaluation Group (TEG), redesignated from the 79th TEG on November 20, 1998, is made up of seven squadrons, two direct-reporting detachments, five squadron detachments and five operating locations at 15 stateside bases. It was activated on December 1, 1991 and replaced the 4443rd Test and Evaluation Group. The 53rd TEG is responsible for the overall management of the 53rd Wing's flying activities at both Eglin and Nellis Air Force Bases. Aircraft assigned to the group include test-configured F-15 Eagles and F-15E Strike Eagles. The 53rd also operates F-16, A-10, HH-60 and F-117 aircraft. The two flying units operating F-15s are the 85th Test and Evaluation Squadron at Eglin AFB and the 422nd Test and Evaluation Squadron at Nellis AFB. Both wear OT tail codes.

F-15C 83-0039 is seen on May 8, 1986 marked as the Tactical Air Warfare Center Flagship. (Ray Leader)

85th TEST AND EVALUATION SQUADRON

The 85th Test and Evaluation Squadron activated on December 1, 1991, is responsible for conducting OT&E on F-15, F-15E and F-16 fighter weapon systems and developing and testing tactics, techniques, and concepts. Specifically, the squadron executes operational testing and evaluation of the newest air-to-ground munitions, air-to-air missiles, reconnaissance systems, electronic warfare systems, and associated subcomponents and avionics. The aircraft kept the OT tail codes and black and white checked tail stripe of its predecessor, the 4485th Test Squadron.

(Brian C. Rogers)

F-15C 85-0126 is seen here at Langley AFB, Virginia on July 8, 1994. (Brian C. Rogers)

This right side view of 85-0126 was taken in October 1994 at Eglin AFB. (Pete Becker/AIR)

59

F-15B 73-0114 is seen here at Eglin AFB on October 22, 1996. (Pete Becker/AIR)

F-15A 74-0124 is seen in June 1989. (David F. Brown)

F-15C 78-0542 is seen here at McConnell AFB on October 4, 1993. (John Cook)

F-15A 77-0064 is seen in May 1991 in markings of the 53rd Wing. (David F. Brown)

F-15C 82-0015 is seen during October 1994 at Eglin AFB, Florida. (Pete Becker/AIR)

F-15C 80-0047 is seen on September 15, 1996 at Luke AFB, Arizona in markings of the 85th TES. (Kevin Patrick)

F-15A 77-0095 is seen in June 1993. (David F. Brown)

F-15E 88-1681 is seen in May 1995 in the markings of the 85th TES. (Alec Fushi)

(Pete Becker)

WARNER ROBINS AIR LOGISTICS CENTER

339TH FLIGHT TEST SQUADRON
Warner Robins Air Logistics center provides Depot level maintenance and world wide logistics management for the F-15 fleet. The ALC is located at Robins AFB, Macon, Georgia. The 339th Flight Test Squadron is responsible for F-15 flight test activities occurring from Robins AFB.

Left: F-15A 77-0068 is seen on April 9, 1986 with RG (Robins Georgia) tail codes. (Kevin Patrick Collection)

Below: F-15A 77-0068 is seen carrying a WR (Warner-Robins) tail code. (Don Logan Collection)

F-15E 86-0189 is seen in November 1994 carrying the RG tail code. Of note is the 422nd Test & Evaluation Squadron emblem on the left intake. (Pete Becker)

CHAPTER FOUR

Ground Training Units

SHEPPARD AFB, TEXAS, 82nd TRAINING WING

The 82nd Training Weapons Wing (TRW) began operations at Sheppard AFB on July 1, 1993. As a unit of The Air Education and Training Command, the 82nd TRW is responsible for training USAF personnel in many fields including aircraft maintenance. To accomplish this function the wing is divided into seven groups, with the 82nd Training Group being responsible for aircraft maintenance training. The Wing was previously designated the 82nd Flying Training Wing and assigned to Williams AFB, Arizona. It was responsible for Undergraduate Pilot Training at Williams AFB, Arizona. With the closure of Williams AFB, the Wing was inactivated on March 31, 1993 in preparation for its move to Sheppard AFB.

362nd Training Squadron

The 362nd Training Squadron (TRS) is assigned to the 82nd Training Group (TRG) and uses various aircraft to train USAF aircraft maintenance personnel. For F-15 training, surplus F-15 aircraft designated GF-15A and GF-15B are assigned to the 82nd TRG and used for ground training. All Sheppard AFB training aircraft are marked with ST (Sheppard Training) Tail Codes. The GF-15s assigned for training are no longer flyable.

The following are the 13 GF-15As and 5 GF-15Bs assigned to the 82nd TRG:

GF-15A; 72-0115, 74-0119, 76-0008, 76-0022, 76-0054, 76-0067, 76-0079, 76-0083, 76-0110, 77-0085, 77-0095, 77-0125, and 77-0150
GF-15B; 74-0142, 76-0135, 77-0154, 77-0156, and 77-0157

GF-15A 74-0119 ground maintenance Trainer, one of thirteen GF-15As at Sheppard AFB is seen on February 11, 1999. (Kevin Patrick)

GF-15A 76-0067 is seen here at Sheppard AFB on March 1, 1999. In addition to the thirteen GF-15As, there are five GF-15Bs assigned to the 82nd Training Group at Sheppard AFB. (Kevin Patrick)

GF-15A 76-0083 is seen here at Sheppard AFB on March 1, 1999. (Kevin Patrick)

CHAPTER FIVE

Operational USAF F-15 Units

The first combat Wing to receive the F-15A was the Tactical Air Command's (TAC) 1st Tactical Fighter Wing at Langley AFB, Virginia. The Wing had previously flown F-4Es at MacDill AFB, Florida, and was moved without personnel or equipment to Langley AFB during summer 1975. They received their first F-15 on December 17, 1975. In early 1977, the first F-15s assigned in Europe began deliveries to the 36th TFW based at Bitburg, Germany. The 49th TFW at Holloman AFB in New Mexico, began to receive the Eagle in early 1977, becoming the second operational TAC F-15 Wing. Also, in 1977, the 57th Fighter Weapons Wing at Nellis AFB, Nevada received its first F-15A/B Eagles. These were assigned to the 433rd Fighter Weapons Squadron.

In 1978, Eagles arrived at Soesterberg in the Netherlands and were assigned to the 32nd Tactical Fighter Squadron. Also in 1978, the 33rd TFW at Eglin AFB, Florida received F-15A/B Eagles. Additional units continued to receive F-15s throughout the late 1970s and early 1980s.

The last of the 409 production F-15As and Bs that were built for the USAF were delivered to Fighter-Interceptor Squadrons (FIS) then assigned to TAC for the air defense of the United States. The FISs had been assigned to the Air Defense Command (ADC) (known in its last years as the Aerospace Defense Command), which had turned over its assets to Air Defense Tactical Air Command (ADTAC), a unit of TAC, in October 1979. A further organization change in 1985 resulted in ADTAC becoming the First Air Force. Four FIS squadrons (5th, 48th, 57th, and 318th) received the F-15A/B for use in the interceptor role, replacing the Convair F-106 Delta Dart. The active Air Force's Eagle interceptor squadrons were inactivated during the early 1990s, and their aircraft were reassigned to Air National Guard units.

The F-15C/D, with its improvements, began delivery to the USAF in 1979. The 18th TFW at Kadena AB Okinawa became the first unit in the Pacific Theater to operate F-15s, receiving its F-15Cs during 1979. The 32nd Tactical Fighter Squadron at Soesterberg AB was the first USAFE unit to operate the F-15C/D, replacing its F-15As and Bs during 1980. The 57th FIS at Keflavik in Iceland, and a second squadron in the Alaskan Air Command (21st TFW, 54th TFS) also received F-15Cs. The F-15C/D replaced the F-15A/Bs in all USAF units, with the exception of the 49th TFW at Holloman AFB. A total of 408 F-15Cs and 59 F-15Ds were delivered to the

F-15B 73-0108 (TAC 1), Tactical Air Commands first F-15, is seen here at its delivery ceremony to the 58th Tactical Training Wing on November 14, 1974 at Luke AFB, Arizona. (Dennis R. Jenkins)

USAF. Many of the F-15A/Bs replaced by the F-15C/Ds were passed along to Air National Guard units.

In the late 1980s, the F-15E Strike Eagle dual role version of the F-15 began to be assigned to Air Force squadrons. The first unit to get the F-15E was the 461st Tactical Fighter Training Squadron (TFTS), part of the 405th Tactical Training Wing at Luke AFB, Arizona. The 461st TFTS mission was to train the crews for the F-15Es. The first operational F-15E unit was the 336th Tactical Fighter Squadron, 4th Tactical Fighter Wing (now just the 4th Fighter Wing), stationed at Seymour-Johnson AFB, North Carolina.

From spring 1984 through early 1998, F-15 fighter-interceptor pilot training was accomplished by the 325th Fighter Wing at Tyndall AFB. In preparation for F-22 pilot training, F-15 fighter-interceptor pilot training has begun to relocate to the 173rd Fighter Wing, Oregon Air National Guard, Kingsley Field.

Presently ACC has three combat wings operating F-15s. The 1st Fighter Wing at Langley AFB, Virginia operates three squadrons of air superiority F-15C/Ds, the 4th Fighter Wing at Seymour-Johnson, North Carolina operates 4 squadrons of F-15E Strike Eagles, and the 366th Wing at Mountain Home AFB, Idaho operates one squadron of F-15C/Ds and one squadron of F-15Es. PACAF has two F-15 units; the 18th Wing at Kadena Okinawa operates three squadrons of air superiority F-15C/Ds, and the 3rd Wing at Elmendorf AFB, Alaska operates two squadrons of F-15C/Ds and one squadron of F-15Es. The 48th Fighter Wing at RAF Lakenheath, England, the last F-15 unit in Europe, operates two squadrons of F-15Es and one squadron of F-15C/Ds.

AIR DEFENSE UNITS

AIR DEFENSE TACTICAL AIR COMMAND (ADTAC)
1st AIR FORCE

The 1st AF was responsible for the air defense of the continental United States and Canada. It was a descendant of the Air Defense

Command (ADC). On October 1, 1979, ADC transferred its responsibilities to TAC, with TAC forming the Air Defense Tactical Air Command (ADTAC) at the former ADC headquarters in Colorado Springs, Colorado. The ADTAC headquarters were transferred to Langley AFB, Virginia, on June 1, 1981. On December 6, 1985 the organization was redesignated the 1st Air Force. 1st Air Force resources included operations control centers, radar warning systems, and fighter aircraft used to conduct peacetime air sovereignty and wartime air defense missions. 1st Air Force acquired the F-15A/B in the early 1980s for the air defense role. In addition the air defense version of the F-16 was assigned to First Air Force. Since 1989, the 1st Air Force has supported the US Customs Service and a variety of law enforcement agencies in their efforts to track and identify air traffic suspected of carrying illegal drugs into the United States.

Four F-15 equipped Fighter Interceptor Squadrons (FIS) were originally assigned to the First Air Force, with the three assigned to the Continental U.S. using early F-15A/B models in the air interceptor role. The air defense role was subsequently transferred to the F-16, and the F-15s were redistributed to ANG squadrons. The First Air Force is now an Air National Guard unit responsible for the ANG's Air Defense Units.

F-15B 76-0126 in 48th Fighter Interceptor Squadron markings carries 1st AF on the tail, indicating it is the 1st Air Force Flagship. (David F. Brown)

57th FIGHTER INTERCEPTOR SQUADRON
BLACK KNIGHTS
KEFLAVIK NAVAL STATION, ICELAND

Air defense of Iceland under the operational control of US Atlantic Command. The 57th Fighter Interceptor Squadron, Keflavik NS, Iceland was the Air Forces Iceland's combat element. From Keflavik, fighters of the 57th FIS regularly scrambled to intercept Soviet Bear bombers and reconnaissance aircraft in the area known as the Greenland, Iceland, United Kingdom gap. The 57th FIS was initially equipped with F-89s, and converted to the F-102A in 1962. In 1972 they converted to the F-4C, replacing the C model phantoms with F-4Es in 1978. The 57th converted to F-15C/Ds from F-4E beginning in July 1985, with the last F-4E leaving in February 1986. The C/D model was chosen instead of the A/B since the squadron, flying over frigid ocean and ice covered wastelands, needed the extra range and endurance provided by the conformal fuel tanks carried by the C and D model. On May 31, 1993, as part of an Air Force restructuring initiative to retain its most illustrious combat units, Air Forces Iceland was inactivated and replaced by the 35th Wing; however, one year later, on October 1, 1994, the 35th Wing was redesignated 85th Wing. On May 31, 1993, the 57th FIS was redesignated the 57th Fighter Squadron. The 57th Fighter Squadron was inactivated on March 1, 1995. In early 1995 the air intercept mission from Iceland was taken over by ACC and Air National Guard fighter units on a rotational basis.

Two F-15Cs of the 57th Fighter Interceptor Squadron escort an Soviet Tu-142 Bear bomber. (Boeing)

Since the F-102, 57th aircraft markings have always consisted of black and white checkerboard tail markings. The F-102s and F-4Cs rudders were painted with the checkerboard. The F-4Es initially had a checkerboard tail stripe at the top and bottom of the vertical tail fin. The bottom stripe was later deleted, leaving only the top one. The 57th's F-15s had the 57th's FIS black checkerboard tail stripe on the outside top of both fins and a map of Iceland with an Eagle's head in the center on the inside. The Eagles wore an IS tail code derived from the Icelandic spelling of Iceland.

Below: F-15C 80-0035 is seen on September 18, 1993 marked as the 35th Wing Flagship and carries the black and white checkerboard tail stripe. (Jim Goodall)
Right: (Paul Hart)

F-15C 80-0038 is seen on October 19, 1988 at Tyndall AFB while taking part in William Tell 1988. (Pete Wilson)

F-15C 80-0046 is seen on September 18, 1993. The 57th FIS was one of the few F-15 Air Superiority units to use the Conformal Fuel Tanks (CFT) on their aircraft. They were carried to increase the range and flight duration safety margin on missions over the icy waters of the north Atlantic. (Jerry Geer)

F-15C 80-0050 is seen on October 19, 1988 at Tyndall AFB while taking part in William Tell 1988. (Pete Wilson)

F-15C 80-0048 is seen here at Langley AFB, Virginia on February 17, 1994 also carrying CFTs . (Brian C. Rogers)

48th FIGHTER INTERCEPTOR SQUADRON
LANGLEY AFB, VIRGINIA

48th Fighter Interceptor Squadron, Langley AFB, Virginia received its first F-15A/B on August 14, 1981, replacing the F-106A. The unit retained their blue and white strip and echelon tail markings for a while, but later reverted to standard TAC-style tail codes using LY for LangleY AFB. The squadron officially stood-up with the F-15A/B on April 5, 1982, and was inactivated on September 30, 1991. The squadron's F-15s were transferred to the 131st FW of the Missouri ANG. This squadron received three F-15As modified to carry the ASM-135A anti-satellite (ASAT) missile, although the missile was never operational. When inactivated on December 31, 1991, the aircraft were sent to 131st Tactical Fighter Wing, Missouri ANG.

(Brian C. Rogers)

F-15A 76-0100 is seen on October 23, 1986 marked as the 48th FIS Flagship. (Pete Wilson)

71

F-15B 76-0126 is seen here at Luke AFB, Arizona on November 5, 1988, carrying the LY tail code of the 48th FIS, marked as the 1st AF Flagship. (Kevin Patrick)

F-15A 76-0088 is seen here at Luke AFB on January 13, 1990 carrying LY tail codes and marked as the 48th FIS Flagship. (Kevin Patrick)

In this December 10, 1989 photo, F-15A 76-0108 is seen here at Nellis AFB carrying LY tail codes also marked as the 48th FIS Flagship. (Kevin Patrick)

Right: F-15A 76-0088 is seen carrying the tail markings of the 48th FIS. (David F. Brown)

Below: F-15A 76-0090 is seen carrying the tail markings of the 48th FIS including, on the left intake, the early Air Defense Command version of the 48th FIS patch. (David F. Brown)

F-15A 76-0115 is seen on December 10, 1989 at Nellis AFB carrying the LY tail code of the 48th FIS. (Brian C. Rogers)

F-15A 77-0137 is seen on December 10, 1989 at Luke AFB in 48th FIS markings. (Douglas Slowiak/Vortex Photo Graphics)

F-15A 77-0145 is seen on December 10, 1989 at Luke AFB carrying the LY tail code of the 48th FIS including, on the left intake, the Tactical Air Command version of the 48th FIS patch. (Brian C. Rogers)

318th FIGHTER INTERCEPTOR SQUADRON
GREEN DRAGONS
McCHORD AFB, WASHINGTON

The 318th FIS at McChord AFB, Washington, traded its F-106As for F-15A/Bs in November 1983. The squadron was inactivated on 7th December 1989 and its aircraft transferred to the 142nd FIG of the Portland ANG. Like the 48th, this squadron received three F-15As wired for the Vought ASAT missile, although no missiles were ever operational. Originally the squadron retained their colorful eight pointed two-tone blue star on the vertical stabilizers. Later in their career the 318th adopted TC (for TaComa, a nearby city in Washington State) tail codes and a two-tone blue fin stripe with an eight pointed star in its center.

(Boeing)

Aircraft of the 318th FIS stood satellite alert at Castle AFB, California as Detachment I of the 318th FIS. Detachment I was activated on October 1, 1981, and closed down in 1989 shortly before the 318th was inactivated.

F-15A 76-0111 is seen here at Nellis AFB in October 1989 carrying the 318th TC tail code and marked as the 318th Commanders Flagship. (David F. Brown)

F-15A 76-0111 is seen here at Nellis AFB in October 1988 carrying the 318th eight pointed blue star on the tail and marked as the 318th Commanders Flagship. (Doug Remington)

F-15A 76-0057 is seen here at McChord AFB during September 1986 taxiing back from a mission. (Jim Goodall)

F-15A 76-0100 is seen here at Nellis AFB on October 6, 1988 carrying the 318th TC tail code. (Brian C. Rogers)

F-15A 76-0093 is seen on November 5, 1988 at Luke AFB carrying the TC tail code of the 318th FIS. (Kevin Patrick)

77

5th FIGHTER INTERCEPTOR SQUADRON
SPITTIN' KITTENS
MINOT AFB, NORTH DAKOTA

The 5th Fighter Interceptor Squadron (FIS) received its first F-15A/B in April 1985, replacing the General Dynamics F-106 Delta Dart. The Minot AFB-based unit retained the colorful yellow lightning bolt tail motif used on their F-106s. The unit operated F-15s for barely three years, being inactivated on July 1, 1988. The 5th FIS aircraft were sent to the 102nd Fighter Interceptor Group, Massachusetts ANG. The defense mission for the northern border was transferred to the 119th FIG of the North Dakota Air National Guard, Fargo, North Dakota.

F-15A 76-0015 is seen marked as the 5th FIS Commanders Flagship. (David F. Brown)

F-15A 76-0011 is seen here at Minot AFB in August 1987. (Alec Fushi)

F-15A 76-0017 is also seen at Minot AFB in August 1987. (Alec Fushi)

F-15A 76-0021 is seen in August 1987. (Alec Fushi)

F-15A 76-0058 is seen here at Tyndall AFB during William Tell. (David F. Brown)

TACTICAL AIR COMMAND/AIR COMBAT COMMAND UNITS

AIR COMBAT COMMAND

The F-15s entered operational service as a Tactical Air Command (TAC) resource. In recognition of this the first F-15, an F-15B (73-0108) was named TAC - 1. The continental United States based F-15 remained in TAC until the creation of the Air Combat Command (ACC) which occurred on June 1, 1992. This resulted largely from dramatic changes In the international arena. With the collapse of the former Soviet Union, and the end of the Cold War, the Air Force began to reconsider the long-standing distinction between the Strategic Air Command (SAC) and the Tactical Air Command. SAC had become almost totally linked to the notion of nuclear deterrence, while TAC operations were perceived as a cooperative mission with the Air Force working with ground and naval forces. The distinction, however, did not lend itself to a limited conflict. During the war in Southeast Asia, strategic B-52 bombers performed 'tactical' missions, while tactical fighter aircraft carried out strategic bombing deep into enemy territory. In addition, tactical aircraft in Europe routinely sat nuclear alert. Operation Desert Storm in early 1991 further blurred the distinction between the two terms. Consequently, as senior Air Force officials sought to reexamine roles and missions, it was determined that all operational USAF combat aircraft based in the continental United States would be assigned to the new ACC.

Another significant change resulted from an overhaul of flying training responsibilities. Prior to the activation of ACC, TAC was responsible for aircraft specific aircrew training, including initial weapon system and continuation training. On July 1, 1993, the 58th and 325th Fighter Wings, the F-16 and F-15 training units, were transferred from ACC to the Air Education and Training Command (AETC). Concurrently, the unit's home bases, Luke AFB and Tyndall AFB, also moved from ACC to AETC ownership.

(Boeing)

EGLIN AFB, FLORIDA, 33rd FIGHTER WING

58th FS, 59th FS, and 60th FS

The 33rd Fighter Wing presently operates two F-15C/D squadrons, the 58th FS and 60th FS from Eglin AFB, Florida. 33rd aircraft wear the EG (EGlin AFB) tail code and a tail stripe of the squadron color. The 33rd traces its history to the 33rd Pursuit Group, which was activated at Mitchel Field, New York, on January 15, 1941. The Wing patch represents its motto, "Fire From The Clouds." The sword and fire symbolize the righting of a wrong and retribution to the enemy.

The 33rd Fighter Wing has a long history as a combat unit. It fought in campaigns during World War II in North Africa, the Mediterranean, China, Burma and India flying P-38, P-40 and P-47 aircraft. The 33rd was involved in the occupation of Germany after the war and was later assigned to New Mexico flying the P-51. In 1947, the 33rd converted to the F-84 jet. The following year, the unit moved to Otis Air Force Base, Massachusetts, and flew the F-86 until inactivated in 1952. In 1956, it was again activated, assigned F-89s and F-94s, and operated as part of the Eastern Air Defense Force. It was inactivated in 1957. The 33rd Fighter Wing was reactivated at Eglin Air Force Base, Florida, on April 1, 1965, and equipped with the F-4 Phantom II. During the war in Southeast Asia, the wing trained, equipped, and deployed eight combat squadrons to that area.

(Boeing)

(Brian C. Rogers)

(Don Logan)

The 33rd Fighter Wing received its first F-15B (77-0156) for maintenance training on September 21, 1978. The official F-15 arrival ceremony was held on December 15, 1978, with the final F-15A/B arriving on June 21, 1979. The Wing began receiving F-15C/Ds on July 3, 1979 with the delivery of the first production F-15C (78-0470). However, the Wing was simply acting as a staging area for the 18th TFW at Kadena, and was not considered operationally capable. After the F-15C/Ds were sent to Kadena under Operation Ready Eagle III, the 33rd began receiving F-15A/Bs from the 32nd TFS at Soesterberg, The Netherlands. The Wing finally began receiving its own F-15C/Ds on February 23, 1983.

While participating in Desert Storm, the 33rd Fighter Wing had 16 aerial kills including the first three kills of the war. The 33rd's combat achievement accounted for nearly 50 percent of the total Coalition air force's kills. As a leader in the air-to-air combat arena, the 33rd Fighter Wing was also the first unit to carry the AIM-120 Advanced Medium Range Air-to-Air Missile (AMRAAM). The Wing declared Initial Operational Capability for the AIM-120 in September, 1991.

The 33rd Tactical Fighter Wing was redesignated as the 33rd Fighter Wing on October 1, 1991. The 58th, 59th, and 60th Tactical Fighter Squadrons were redesignated as Fighter Squadrons on November 1, 1991. They were transferred to the operational control of the 33rd Operations Group on December 31, 1991. The 33rd Fighter Wing presently has two flying units – the 58th, and the 60th Fighter Squadrons. The third flying squadron, the 59th Fighter Squadron ceased flying operations on December 12, 1997.

F-15A 75-0033 is seen in July 1983 marked as the 33rd TFW Flagship. (David F. Brown)

F-15A 74-0133 is seen on June 15, 1984 also marked as the 33rd TFW Flagship. (David F. Brown)

F-15C 85-0102 marked as the 33rd TFW Flagship is seen here in May 1991. (David F. Brown)

F-15C 84-0021 is seen in December 1996 marked as the 33rd FW Flagship. (Alec Fushi)

F-15C 83-0017 is seen in February 1987 marked as the 33rd TFW Flagship. (Don Logan Collection)

Marked as the 33rd TFW Flagship, F-15C 78-0533 is seen on December 8, 1980. (Brian C. Rogers)

F-15C 84-0030 is seen on January 31, 1994 marked as the 33rd OG (Operations Group) Flagship. (Nate Leong)

75-0033 is seen on February 1, 1994 marked as the 33rd OSS (Operations Support Squadron) Flagship. (Nate Leong)

F-15C 84-0003 is seen in November 1986 with William Tell markings. (Don Logan Collection)

F-15C 86-0169 is seen in June 1992 with reverse tail code and serial number markings. (David F. Brown)

F-15C 85-0121 is seen in October 1994 with William Tell markings. (David F. Brown)

F-15C 85-0122 is seen in October 1994 with William Tell markings. (David F. Brown)

58th FIGHTER SQUADRON
GORILLAS

The 58th Pursuit Squadron was formed on November 20, 1940 and activated along with its sister squadrons the 59th and 60th, as part of the 33rd Pursuit Group on January 15, 1941 at Mitchel Field, New York. The 58th was responsible for the air defense of the United States at different periods on both coasts until October 1942. The 58th was involved in air combat operations in the Mediterranean Theater, from November 1942 to February 1944, and in the China Burma Campaign from April 1944 to August 1945. Following World War II, the 58th was inactivated then activated again a number of times with basing in Germany; at Roswell, New Mexico; and at Otis Air Force Base, Massachusetts. The unit's final activation occurred at Eglin Air Force Base, in 1970 equipped with the F-4E. The 58th saw combat in Southeast Asia while deployed to Udorn, Thailand in 1972.

The 58th TFS received their first F-15A on 15th March 1979, achieving their IOC on May 23, 1979. 58th aircraft wear the EG (EGlin AFB) tail code and a blue tail stripe. The 58th was selected as the first Air Force fighter squadron to be equipped with the AIM-120A AMRAAM weapons system. In December 1989, the squadron was called upon to lead the way for Operation JUST CAUSE by flying initial force protection sorties for invasion missions. The squadron converted to the F-15C/D prior to deploying to Tabuk Air Base, Saudi Arabia, in August 1990 as part of 'Desert Shield'. The 58th scored 13 air-to-air kills during the Gulf War, the highest total from any single squadron. Beginning in February 1991 the squadron became operational with the AIM-120A, and although several hundred sorties were flown with the weapon during the war, none were fired in anger.

F-15C 84-0100 is seen in December 1996 marked as the 58th FS Flagship. (Alec Fushi)

F-15A 76-0058 is seen on August 20, 1984 marked as the 58th FS Flagship. (Keith Svendsen)

F-15C 85-0100 is seen on January 16, 1995 marked as the 58th FS Flagship. (Don Logan Collection)

F-15A 75-0068 is seen in January 1984. (David F. Brown)

F-15A 77-0103 is seen here at Nellis AFB on July 16, 1979. (Brian C. Rogers)

F-15C 85-0101 is seen on June 2, 1994 with a single green star below the windshield, indicating one kill in Operation Desert Storm. (Tom Kaminski)

F-15C 85-0102 is seen on February 1, 1994 with three green stars below the windshield, indicating three kills in Operation Desert Storm. (Nate Leong)

F-15C 85-0105 is seen in June 2, 1994 with two green stars below the windshield, indicating two kills in Operation Desert Storm. (Alec Fushi)

F-15C 85-0119 is seen on June 2, 1994 with a single green star below the windshield, indicating one kill in Operation Desert Storm. (Ben Knowles)

F-15D 85-0129 is seen here at Elgin AFB in December 1996. (Alec Fushi)

59th FIGHTER SQUADRON
PROUD LIONS

Like the 58th Fighter Squadron the 59th began its life as the 59th Pursuit Squadron being formed on November 20, 1940 and activated as part of the 33rd Pursuit Group on January 15, 1941 at Mitchel Field, New York. The 59th was involved in air combat operations in the Mediterranean Theater, from November 1942 to February 1944, and in the China Burma Campaign from April 1944 to August 1945. Following World War II, the 58th was inactivated then activated again a number of times. On 16th March 1970 the 59th FIS was redesignated the 59th TFS, and was assigned to the 33rd TFW on September 1, 1970. During 1979, the squadron began transitioning from the F-4 to the F-15C/D. The 59th TFS also deployed to Operation 'Desert Storm', scoring two air-to-air kills during the conflict. Like the other 33rd FW squadrons, the 59th FS was based at Tabuk, Saudi Arabia. The 59th FS was equipped with the F-15A and B Eagle in 1978 and later the F-15C and D. 59th aircraft wore the EG (EGlin AFB) tail code and a yellow tail stripe. On December 12, 1997 the squadron stood down from flying duties.

F-15C 84-0013 is seen here at Eglin AFB in December 1996 marked as the 59th FS Flagship. (Alec Fushi)

Right: F-15A 76-0059 is seen in July 1981 marked as the 59th TFS Flagship. (Don Logan Collection)

Below: F-15C 86-0159 is seen in April 1990 marked as the 59th TFS Flagship. Later on March 24, 1999, while assigned to the 493rd FS, 48th FW, 86-0159 was credited with downing a Serbian MiG-29. (Alec Fushi)

F-15A 77-0150 is seen here at Davis-Monathan AFB on February 2, 1980. (Brian C. Rogers)

F-15A 74-0115 is seen here at Offutt AFB in May 1982. (George Cockle)

F-15D 76-0120 is seen here at Davis-Monathan AFB on August 27, 1983. (David F Brown Collection)

F-15C 78-0544 is seen here at Davis-Monathan AFB on December 8, 1980. (Brian C. Rogers)

F-15C 80-0010 is seen here at Eglin AFB in December 1996. (Alec Fushi)

F-15C 80-0028 is also seen at Eglin AFB in December 1996. (Alec Fushi)

F-15C 84-0016 is seen here at Langley AFB on March 8, 1994. (Brian C. Rogers)

60th FIGHTER SQUADRON
FIGHTING CROWS

Activated in 1940 at Mitchel Field, NY as the 60th Pursuit Squadron, the Unit was attached to the 33rd Pursuit Group on January 15, 1941. Redesignated as the 60th Fighter Squadron – the Fighting Crows – on May 15, 1942, the Unit was responsible for the continual mission of air defense of the United States until October 1942. In late 1942, the 60th joined the United States effort in World War II by participating in combat operations in the Mediterranean Theater and the China, Burma, India Theater. The squadron's assignments followed its sister squadrons, the 58th and 59th

Between June 15, 1979 and April 16, 1980 the 60th TFS trained 55 pilots and processed 54 F-15C/Ds for the 18th TFW as part of the Kadena Ready Eagle operation. The squadron converted back to the F-15A/B before receiving their own F-15C/Ds. 60th aircraft wear the EG (EGlin AFB) tail code and a red tail stripe. The 60th made its first combat deployment since World War II when it sent ten F-15s to Grenada in support of Operation Urgent Fury, the rescue of American medical students held in Grenada. The unit flew support missions for Operation Just Cause, the removal of Panamanian dictator Manuel Noriega from Panama. The squadron scored a single air-to-air kill during Operation Desert Storm.

F-15C 80-0020 is seen here at Eglin AFB in December 1996 over the runway, just before touchdown. (Tony Cassanova)

F-15C 85-0121 is seen on February 1, 1994 at Eglin AFB marked as the 60th FS Flagship. (Nate Leong)

F-15C 84-0002 is seen in February 1987 at Eglin AFB, marked as the 60th TFS Flagship. (Don Logan Collection)

F-15A 74-0120 is seen here at Luke AFB in February 1985. (Kevin Patrick)

(Boeing)

F-15A 75-0037 is seen here at Nellis AFB on October 16, 1982. (Brian C. Rogers)

F-15A 74-0123 is seen here at Offutt AFB in July 1982. (George Cockle)

F-15C 85-0093 is seen on October 24, 1986. (Nate Leong)

F-15D 85-0130 is seen here at Langley AFB in June 15, 1993. (Brian C. Rogers)

F-15C 85-0121 is seen here at Eglin AFB in December 1996. (Alec Fushi)

F-15D 85-0134 is also seen at Eglin AFB in December 1996. (Alec Fushi)

HOLLOMAN AFB, NEW MEXICO, 49th TACTICAL FIGHTER WING

7th TFS, 8th TFS, and 9th TFS

The 49th Tactical Fighter Wing, stationed at Holloman AFB flew F-15 from early 1977 to September 30, 1991. The HW aircraft wore HO (HOlloman AFB) tail codes and tails stripes corresponding to the assigned squadron color.

Initially designated the 49th Pursuit Group (Interceptor) in 1940, the unit was among the first to deploy from the United States to the Pacific Theater of Operations during World War II. Redesignated the 49th Fighter Group, the Unit played an important role in halting the Japanese advance in the southwest Pacific. By the war's end, the group's pilots had destroyed 678 enemy aircraft, a record surpassing that of any other fighter group in the Pacific Theater. Among the Unit's 43 aces were Lt. Colonel Boyd D. "Buzz" Wagner, the first World War II ace in the Pacific Theater, and Major Richard Bong, whose 40 kills made him America's number one ace (a record that still stands).

On August 19, 1948, the 49th was redesignated the 49th Fighter Wing. From 1946 until 1950, the Wing flew the P-51 and F-80 aircraft as a part of the occupational forces in Japan. Redesignated the 49th Fighter Bomber Wing, the 49th began operations in Korea in June 1950. Following the cessation of hostilities in Korea, the 49th moved from Misawa Air Base, Japan, to Etain-Rouvres Air Base, France. During its time in France, the wing flew the F-84 and then the F-100D. In August 1959, the 49th Tactical Fighter Wing began a nine year stay at Spangdahlem Air Base Germany. The 49th changed to the F-105Ds in 1961 and F-4Ds in 1967.

On July 15, 1968, the 49th began arriving at Holloman AFB, New Mexico. In May, 1972, the 49th deployed with its F-4D aircraft to Takhli Royal Thai Air Base, Thailand, to support combat operations in Southeast Asia as part of Operation Constant Guard. During five months of combat, the Wing did not lose any aircraft. The unit officially closed out its Southwest Asia duty October 6, 1972, being replaced at Takhli by F-111As of the 474th TFW from Nellis AFB.

The 49th FW became a user of late-production F-15A/Bs beginning December 20, 1977, replacing F-4Ds at Holloman AFB, New Mexico. IOC (Initial Operational Capability) was achieved on June 4, 1978. The 7th FS was the first of the Wing's squadrons to receive the F-15. The Wing was the last regular Air Force operator of the F-15A/B, and the only original user not to convert to the newer F-15C/D. The last F-15 departed Holloman on June 5, 1992, ending 14 years of Eagle operations. The 49th subsequently became the Air Force's only operator of the F-117 stealth fighter. The 'HO' tail codes reflect their home base, and most aircraft carried a subdued Eagle motif on the inside of the vertical stabilizers.

The 49th Tactical Fighter Wing was redesignated as the 49th Fighter Wing on October 1, 1991 and was assigned to the newly created Air Combat Command (ACC) on June 1, 1992. The 7th,

(Don Logan)

8th, and 9th Tactical Fighter Squadrons were redesignated as Fighter Squadrons on November 1, 1991 and transferred to operational control of the 49th Operations Group on November 15, 1991. The 7th FS ceased F-15 operations on September 13, 1991; the 8th FS ceased F-15 operations in February 1992. The 9th FS, the 49th Fighter Wing's last F-15 squadron ceased F-15 operations on June 5, 1992.

The last F-15 departed Holloman June 5, 1992, ending 14 years of Eagle operations. On May 9, 1992, four F-117 stealth fighters from Tonopah Test Range, Nevada, arrived at Holloman AFB. This move of the F-117s was completed July 7th, with the last of the personnel and Nighthawks belonging to the 415th, 416th and 417th Fighter Squadrons arriving at Holloman.

(Don Logan)

F-15A 76-0063 is seen here at George AFB in April 1992 marked as the 49th FW Flagship, the "City of Alamogordo". (Don Logan)

F-15A 76-0049 is seen here at Nellis AFB on March 31, 1978 with a modified serial number and a multi-colored tail stripe as the 49th TFW Flagship. (Don Logan)

F-15A 77-0142 is seen here at Norton AFB, California in April 1992 marked as the 49th TFW Flagship. (Mick Roth)

F-15A 77-0081 is seen here at Davis-Monathan AFB in October 1992 marked as the 49th TFW Flagship, the "City of Alamogordo". (Ben Knowles)

F-15A 77-0112 is seen on September 11, 1986, marked as the 12th Air Force Flagship. (Bob Leavitt)

F-15A 77-0106 is seen here at Andrews AFB in January 1989 marked as the 833rd Air Division Flagship. (David F. Brown)

F-15A 77-0078 is seen on October 8, 1984 with the multicolor tail stripe applied for William Tell 1984. (Kevin Patrick)

F-15A 77-0132 is seen on October 23, 1986 with the multicolor tail stripe applied for William Tell 1986. (Kevin Patrick)

F-15A 77-0135 is seen in October 1984 with the multicolor tail stripe and New Mexico Sun emblem applied for William Tell 1984. (David F. Brown)

7th TACTICAL FIGHTER SQUADRON
BUNYAPS

The 7th Tactical Fighter Squadron began on January 16, 1941, when the 49th Pursuit Group (Interceptor) was activated at Selfridge Field, Michigan. During that activation, the 7th, 8th, and 9th Pursuit Squadron (Interceptor) were assigned to the Group and remained a part of the 49th throughout its history. The 7th TFS was the first 49th TFW squadron to receive the F-15 beginning its transition to the F-15 in 1977. 7th squadron aircraft wore the HO tail code and a blue tail stripe. For a short time, during 1979 and 1980, a blue and white checkerboard tail stripe was used. The unit's last F-15 flight occurred on September 13, 1991, with the squadron inactivating on September 30, 1991. It has since been reactivated flying the F-117.

(Jim Goodall)

F-15A 77-0137 is seen here at Luke AFB on July 11, 1988 marked as the 7th TFS/AMU Flagship. (Kevin Patrick)

F-15A 76-0084 is seen here at Nellis AFB on May 14, 1978. (Don Logan)

F-15A 76-0097 is seen here at Nellis AFB on May 14, 1978. (Don Logan)

F-15A 77-0104 is seen here at Nellis AFB in August 1990. (David F. Brown)

F-15B 76-0133 is seen on March 20, 1978. As evidenced by the white area on the fuselage side aft of the moveable intake, the M61A1 20mm cannon has been removed. (Ben Knowles)

F-15A 77-0133 is seen here at on May 3, 1980 with the blue and white checked tail stripe used by the 7th TFS during 1980. (Doug Remington)

F-15B 77-0159 is seen here at Nellis AFB in May 1980. (Tom Brewer)

8th TACTICAL FIGHTER SQUADRON
BLACK SHEEP

Like the 7th TFS, the 8th Tactical Fighter Squadron's story began on January 16, 1941, when the 49th Pursuit Group (Interceptor) was activated at Selfridge Field, Michigan. During 1977 the 8th TFS began its transition to the F-15. In June 1978, the transition to the F-15 Eagle was completed. The squadron operated the F-15 until February 1992. In August 1992, the 8th Fighter Squadron began flying the AT-38B Talon. Its mission was to train new Air Force pilots, fresh out of undergraduate pilot training, the skills required for aerial combat. 7th squadron aircraft wore the HO tail code and a yellow tail stripe. On July 30, 1993, the 8th Fighter Squadron transitioned to the F-117A Stealth Fighter.

F-15A 77-0118 is seen here at Luke AFB in May 1981 marked as the 8th TFS/AMU Flagship. (Ben Knowles)

F-15A 76-0063 is seen here at Luke AFB on December 17, 1989 marked as the 8th TFS/AMU Flagship. (Kevin Patrick)

Right: F-15A 77-0118 is seen with a modified serial number and a yellow tail stripe as the 8th TFS Flagship. (Don Logan Collection)

Below: F-15A 77-0128 is seen here at Luke AFB on August 8, 1986. (Douglas Slowiak/ Vortex Photo Graphics)

F-15A 76-0113 is seen here at Davis-Monathan AFB on January 10, 1979. (Brian C. Rogers)

F-15A 76-0093 is seen here at Davis-Monathan AFB on January 11, 1979. (Brian C. Rogers)

F-15A 77-0145 is seen here at Luke AFB on April 19, 1986. (Douglas Slowiak/Vortex Photo Graphics)

F-15B 77-0168 is seen here at Langley AFB on September 1, 1987. (Don Logan Collection)

9th TACTICAL FIGHTER SQUADRON
IRON KNIGHTS

The 9th Tactical Fighter Squadron was activated as the 9th Pursuit Squadron (Interceptor) on January 15, 1941 at Selfridge Field, Michigan. Redesignated as the 9th Fighter Squadron on May 15, 1942, and was assigned to the 49th Fighter Group, and remained with the 49th throughout its history. The unit started receiving its F-15A and B Eagles during 1978. 9th squadron aircraft wore the HO tail code and a red tail stripe. On June 5, 1992 the 9th ceased F-15 operations. the 9th TFS flew F-4Es training German Air Force aircrews until June 1993. The 9th gave up the training role and F-4Es were replaced by F-117s in July 1993.

F-15A 77-0071 is seen here at Luke AFB on September 10, 1988 marked as the 9th TFS/AMU Flagship. (Don Logan Collection)

F-15A 77-0119 is seen here at Tyndall AFB on January 16, 1991 marked as the 9th TFS/AMU Flagship. (Barry Roop)

F-15A 77-0066 is seen here at Williams AFB on January 10, 1979. (Brian C. Rogers)

F-15A 77-0091 is seen here at Luke AFB in November 1984. (Kevin Patrick)

F-15A 77-0128 is seen on June 1, 1988. (Craig Kaston)

LANGLEY AFB, VIRGINIA, 1st FIGHTER WING

27th FS, 71st FS, and 94th FS

On March 14, 1974, the Air Force announced plans to station the first operational F-15 wing at Langley AFB, Virginia. Langley was chosen due to its heritage and ideal location for TAC's secondary air defense mission. On June 6, 1975, Tactical Air Command directed 9th Air Force to move the 1st FW, and its associate squadrons, from MacDill AFB to Langley. Although the designation of the unit moved, the majority of MacDill personnel remained in place, and served under the newly designated 56th TFW. The next six months spent preparing for the arrival of the F-15, and on December 18, 1975 the Wing's first F-15B arrived. Official welcoming ceremonies were held on January 9, 1976, when the first F-15A arrived. 1st FW aircraft wear FF (First Fighter) tail codes and tails stripes corresponding to the assigned squadron color.

After achieving their Initial Operational Capability (IOC), the 1st FW embarked on Operation 'Ready Eagle' to help prepare the 36th TFW at Bitburg AB for their reception of the F-15. By September 23, 1977, the Wing provided Bitburg with 88 operational ready pilots, 522 maintenance specialists, and later trained an additional 1,100 maintenance personnel at Bitburg. The Units changed from F-15A/Bs to F-15C/Ds in the early 1980s.

Beginning on August 7, 1990, the 27th and 71st Tactical Fighter Squadrons deployed 48 aircraft to Saudi Arabia in support of Operation Desert Shield. By 16th January 1991 the Wing had amassed 4,207 sorties. On January 17, 1991, 16 1st TFW F-15s departed King Abdul Aziz Air Base and headed toward Iraq to participate in the opening round of Operation Desert Storm. On 8th March 1991, the 1st TFW returned to Langley from Saudi Arabia. The end of the Gulf War did not bring an end to 1st FW support in Southwest Asia, however, and the 1st FW provided aircraft six months per year to monitor the southern no-fly zone under Operation Southern Watch. In 1996 the 1st FW began participating in Airpower

(David F. Brown)

F-15C 83-0033 is seen here at Langley AFB on July 11, 1988 marked as the 1st FW Flagship. (Don Logan)

Expeditionary Force (AEF) deployments to the Middle East, while continuing to provide forces for Southern Watch at Dhahran and other Persian Gulf area bases. Elements of the 1st FW led the first such AEF deployment to Jordan in spring 1996

On October 1, 1991, the 1st Tactical Fighter Wing was redesignated 1st Fighter Wing The 27th, 71st, and 94th Tactical Fighter Squadrons were transferred to operational control of the 1st Operations Group on October 1, 1991 and on November 1, 1991, redesignated as Fighter Squadrons on November 1, 1991.

A four-ship of 1st Tactical Fighter Wing Flagships over-flies Langley AFB (Boeing)

Above: F-15C 74-0100 is seen on October 21, 1977 marked as the 1st TFW Flagship. (Don Logan)

Right: F-15C 82-0011 is seen here at Langley AFB on July 11, 1988 marked as the 1st TFW Flagship. (Don Logan Collection)

Below: F-15D 80-0060 is seen on May 24, 1986 marked as the 1st TFW Flagship. (Douglas Slowiak/Vortex Photo Graphics)

F-15C 81-0038 is seen here at Langley AFB on July 11, 1988 marked as the 1st OG (Operations Group). (Jim Geer)

F-15C 83-0018 is seen here at Langley AFB October 5, 1992 marked for William Tell 1992. (Don Logan)

F-15C 81-0037 is seen here at Luke AFB on October 25, 1992, having just returned from rotation to the Middle East, marked with both 71st and 94th FS emblems and combination tail stripe. (Kevin Patrick)

In May 1977, the 1st Tactical Fighter Wing history office discovered reference to a special order, signed in September 1944 by Col. Robert B. Richard who then commanded the 1st Fighter Group. The order stated that "in the 27th Fighter Squadron airplane number 23 should forevermore be named 'Maloney's Pony' in honor of the 1LT Tom Maloney.

With eight accredited aerial victories, Lt. Tom Maloney was the leading ace of his unit – the 27th Fighter Squadron of the 1st Fighter Group. During his ten months of combat in the Mediterranean Theater of Operations, he had flown the P-38 Lightning on 64 combat missions, accumulating 306:59 flying hours. Exploding debris from a German train he had just strafed ripped out the P-38 underside causing him to crash. He survived the crash.

F-15C Eagle, serial number 0023, carries the Maloney's Pony nose art, continuing the tradition.

(Kevin Patrick)

F-15C 82-0023 is seen here at Luke AFB in June 1994 marked as "Maloney's Pony". (Kevin Patrick)

F-15C 81-0023 is seen here at Nellis AFB on December 11, 1998 replacing 82-0023 as "Maloney's Pony". (Brian C. Rogers)

27th FIGHTER SQUADRON
FIGHTIN' EAGLES

The 27th Tactical Fighter Squadron can lay claim to being the oldest fighter squadron in the Air Force, with a continuous history dating to June 15, 1917. It was also the 1st Tactical Fighter Wing's first squadron to equip with the F-15 was the 27th Tactical Fighter Squadron, receiving its first F-15 on December 10, 1975. On November 1, 1991 the 27th TFS was redesignated the 27th Fighter Squadron (FS). 27th squadron aircraft wear the FF tail code and a yellow tail stripe.

F-15C 81-0027 is seen in October 1987 marked as the 27th AMU Flagship. 27th AMU is marked on the right vertical stabilizer with 27th TFS on the left stabilizer. (Don Logan Collection)

F-15C 81-0027 is seen here at Luke AFB on November 5, 1988 marked as the 27th TFS Flagship. (Kevin Patrick)

F-15A 74-0084 is seen here at Langley AFB on October 21, 1977. (Ray L. Leader)

F-15A 74-0101 is seen here at Langley AFB in May 3, 1976. (Ray L. Leader)

F-15D 80-0060 is seen in March 1989. (Jerry Geer)

F-15C 81-0038 is seen here at Langley AFB on April 24, 1987. (David F. Brown)

F-15C 82-0016 is seen here at Luke AFB on May 23, 1987. (Douglas Slowiak/Vortex Photo Graphics)

F-15C 83-0033 is seen on August 22, 1997. This aircraft was lost 50 miles off the coast of Virginia on November 27, 1997. The pilot was rescued. (Nate Leong)

71st FIGHTER SQUADRON
THE IRON MEN

The 71st Tactical Fighter Squadron was the last squadron of the 1st Tactical Fighter Wing to equipped with the F-15. On November 1, 1991 the 71st TFS was redesignated the 71st Fighter Squadron (FS). The 71st FS scored a kill on January 17, 1991, the first day of Desert Storm, shooting down an Iraqi Mirage F.1 with an AIM-7. 71st squadron aircraft wear the FF tail code and a red tail stripe.

F-15C 80-0040 is seen in October 1987 marked as the 71st TFS Flagship. 71st AMU is marked on the right vertical stabilizer with 71st TFS on the left stabilizer. (Don Logan Collection)

F-15C 83-0037 is seen here at Langley AFB on April 24, 1987. (David F. Brown)

Four F-15Cs of 71st TFS seen here at Langley AFB in November 1982 await pilots for the day's mission. (David F. Brown Collection)

F-15A 74-0092 is seen taking the runway at Nellis AFB in the early morning sun in October 1978. (Don Logan)

F-15A 74-0098 is seen here at Nellis AFB on April 24, 1987. (Don Logan)

F-15A 75-0018 is seen here at Nellis AFB on March 31, 1978. (Don Logan)

F-15C 82-0009 is seen here at Langley AFB in September 1995. (Jerry Geer)

F-15C 83-0041 is seen here at Langley AFB on July 18, 1987. (David F. Brown)

F-15D 83-0047 is seen here at Luke AFB on October 11, 1996. (Norris Graser)

124

94th FIGHTER SQUADRON
HAT IN THE RING or SPADS

The 94th Fighter Squadron had been made famous by Captain Eddie Rickenbacker in the First World War as the 'Hat in the Ring' squadron, and is the second oldest fighter squadron in the Air Force. The 94th FS was the second 1st Tactical Fighter Wing's squadron to receive F-15s. On November 1, 1991 the 94th TFS was redesignated the 94th Fighter Squadron (FS). 94th squadron aircraft wore the FF (First Fighter) tail code and a blue tail stripe.

F-15C 83-0010 is seen in October 1987 marked as the 94th FS Flagship. (Jim Geer)

F-15C 83-0015 is seen here at Langley AFB in April 1985 marked as the 94th TFS Flagship. (David F. Brown)

F-15C 83-0042 is seen in May 1987 marked as the 71st AMU Flagship. (David F. Brown)

F-15A 74-0094 is seen here at Nellis AFB in July 1976. (Don Logan)

F-15A 74-0103 is seen here at Nellis AFB in July 1976. (Don Logan)

F-15C 81-0032 is seen on June 19, 1992. (Norris Graser)

F-15C 81-0037 is seen with a non-standard paint scheme on April 24, 1987. (Don Logan Collection)

F-15C 81-0050 is seen here at Luke AFB on April 5, 1992. (Douglas Slowiak/Vortex Photo Graphics)

F-15C 83-0022 is seen here at Langley AFB on April 24, 1987. (David F. Brown)

F-15C 81-0034 is seen here at Nellis AFB in September 1995. (David F. Brown)

LUKE AFB, ARIZONA

The F-15 Replacement Training Unit (RTU) started at Luke AFB at the beginning of the F15's operational career. During the years F-15 training was occurring at Luke, the F-15s were assigned to three different Wings, with the Wings going through four redesignations. The F-15s started service in the 58th Tactical Fighter Training Wing (TFTW) with a single squadron, the 555th Tactical Fighter Training Squadron (TFTS). The first F-15, a B model (73-0108) arrived on November 14, 1974. The 58th at the time was also training F-4 aircrews for the USAF and F-104 pilots for the German Air Force. The 58th's LA tail code was representative of Luke, Arizona. The 58th TFTW was redesignated the 58th Tactical Training Wing (TTW) on April 1, 1977. On August 29, 1979 the 58th TTW terminated its F-15 training program and transferred its F-15s to the 405th TTW, activated at Luke on the same date. The 58th TTW remained at Luke as an RTU training F-16 pilots. All F-15 training transitioned to the 405th TTW at the same time. F-15A/B and D training continued until the fall of 1991 when the training of F-15 Eagle pilots was relocated to the 325 TTW, Tyndall AFB, Florida. F-15E training with the 405th continued until October 1, 1991 when, because of the new Air Force policy of one base – one Wing, the F-15E Squadrons of the 405th TTW were transferred to the 58th Fighter Wing (FW) picking up the LF tail code used by the 58th FW. On July 1, 1992 the 58th FW transferred from ACC to AETC (Air Education and Training Command). On April 1, 1994, the 58th FW was replaced by the 56th FW. The 56th retained the LF tail code. Sources indicate two possible derivations of LF, either Luke Field or Luke Fighters. F-15E aircrew training continued under the 56th FW until March 1995 when the last F-15E left Luke for the new F-15E Formal Training Unit (FTU), the 333rd FS, which was part of the 4th Fighter Wing at Seymour-Johnson AFB, North Carolina.

(Boeing)

58th TACTICAL FIGHTER TRAINING WING/TACTICAL TRAINING WING

461st TFTS, 550th TFTS, and 555th TFTS

The 58th Tactical Training Wing (TTW) was primarily a replacement training unit (RTU, later called Formal Training Unit or FTU), training fighter aircrews in air-to-air and air to ground operations. The first F-15 began arriving assigned to the 555th TFTS, 58th TFTW in November 1974. The 58th TFTW was redesignated as the 58th TTW on April 1, 1977. Aircraft of the 58th wore an LA tail code. The Wing was the F-15 RTU, responsible for training of F15 pilots for the USAF. On August 29, 1979 the 58th TTW terminated its F-15 training program and transferred its F-15s to the 405th TTW, activated at Luke on the same date. All three F-15 squadrons, the 461st, 550th, and 555th, transitioned to the 405th TTW at the same time. The 58th was subsequently redesignated a Fighter Wing and became the RTU for the F-16.

Right: The 58th TFTW was one of the first TAC Wings to carry Commander's Flagship markings. The rudders on 73-111, similar to U.S. Navy CAG markings, carry the colors of all the Wing's Flying Squadrons. (Mick Roth)

Below: Three F-15Bs and one F-15A, all in Air Superiority Blue, are lined up on the flight line at Luke AFB on March 18, 1975 ready for the day's mission. (Gary Emery)

SPECIAL 58th TFTW PAINT SCHEMES

Operational 58th TFTW F-4 and F-15 aircraft at Luke AFB received high-visibility markings for a short time during 1976. The markings were applied as a test to determine if increased visibility markings would improve safety and training effectiveness for training aircraft. 73-0100 was painted with wide yellow bands around the wings and tails. These stripes were called Sabre stripes because they were similar to those applied to some F-86s. 73- 0103 was painted with alternating white and red invasion stripes around the forward fuselage and on wings. 73-0112, the 12th Air Force Commander's aircraft received similar (but not identical) stripes in white and black.

Four F-15 aircraft were painted in the Ferris Attitude Deceptive scheme designed by aviation artist Keith Ferris. Ferris schemes

were also applied to several F-14s, as well as Air Force and Navy F-4s, F-5s, and other aggressor aircraft. Also a false canopy shape was painted on the bottom of the aircraft. There were several legal concerns over the continued use of the paint scheme by the military, and the USAF repainted the F-15s into Compass Ghost in late-1976. (See page 135.)

The following Federal Standard (FS) colors were used for the Ferris schemes:

Serial	Light	Medium	Dark	Canopy
73-0111	36440	36231	36118	36118
74-0089	36440	36231	36320	36118
74-0110	36622	36440	36231	36231
74-0139	36440	36231	36118	36118

F-15B 73-0111 was the F-15 used as the 58th Tactical Fighter Training Wing's Flagship. (Mick Roth)

Seen here in Air Superiority Blue, F-15B 73-0112 was used as the 12th Air Force Flagship. (Don Logan)

Seen here at Luke AFB, F-15B 73-0112, as the 12th Air Force Flagship, it had black and white D-Day Invasion Stripes added. The 58th TFTW added stripes to some of their F-15s and F-4s in a test to determine which markings made aircraft visible during training missions. (Mick Roth)

F-15A 76-0067 is seen here at Luke AFB on September 3, 1990 marked as the 12th Air Force Flagship. (Douglas Slowiak/Vortex Photo Graphics)

F-15A 76-0102 is seen here at Luke AFB on November 1, 1986 carrying the 550th TFS tail stripe marked as the 832nd Air Division Flagship. (Douglas Slowiak/Vortex Photo Graphics)

F-15A 73-0107 is seen here at Luke AFB on July 6, 1985 carrying the 555th TFS tail stripe marked as the 832nd Air Division Flagship. (Kevin Patrick)

F-15A 77-0141 is seen here at Luke AFB carrying a multi-color tail stripe marked as the 832nd Air Division Flagship. (Mick Roth)

F-15A 76-0046 is seen here at Luke AFB on June 31, 1988 carrying a multi-color tail stripe marked as the 832nd Air Division Flagship. (Mick Roth)

Above: Seen here at Luke AFB in March 1976, F-15A 73-0100 is wearing yellow and black stripes similar to those carried by Korean War F-86 Sabre Jets as part of the 58th TFTW's test. (Mick Roth)

Right: F-15A 73-0103 is seen here at Luke AFB in December 1975 with red and white D-Day Invasion Stripes added. (Jim Rotramel)

Testing of the "Ferris Attitude Deceptive" paint scheme designed by aviation artist Keith Ferris on three F-15As and one F-15B occurred at Luke AFB during 1976. Similar "Ferris" schemes were tested on USAF and USN F-4s, F-5s, and F-15s. F-15B 73-111 is seen in November 1978 still wearing the "Ferris" scheme. (Don Logan)

F-15A 74-0089 is seen here at Luke AFB in May 1976 in the "Ferris" paint scheme. (Jim Rotramel)

F-15A 74-0110 is seen here at Luke AFB in May 1976 in the "Ferris" paint scheme. (Jim Rotramel)

F-15B 74-0139 is seen here at Luke AFB in August 1976 in the "Ferris" paint scheme. (Mick Roth)

405th TACTICAL TRAINING WING

426th TFTS, 461st TFTS, 550th TFTS, and 555th TFTS

The 405th TTW was activated at Luke AFB on August 29, 1979, becoming the RTU for all Air Force F-15A/Bs. The 405th took control of all resources of the 58th Tactical Training Wing, including three Luke F-15 squadrons at the same time, and added the 426th TFTS in January 1981. In 1988 the Wing began transitioning to the Strike Eagle, becoming the RTU for all Air Force F-15Es. By October 1, 1991, when the 405th TTW was inactivated, all F-15 Eagle air superiority/interceptor training had transferred to the 325th TTW at Tyndall AFB. Also on October 1, 1991, the F-15E training squadrons, the 461st FS, 550th FS and 555th FS were transferred to the 58th FW.

Below: F-15A 74-0122 seen at Luke AFB on February 16, 1985 is marked as the 405 Tactical Training Wing (TTW) Flagship. (Kevin Patrick) Right: (Don Logan Collection)

F-15A 76-0089 seen at Luke AFB in March 1983 is wearing the tail stripe of the 426th TFS and marked as the 405 TTW Flagship. In addition the tail stripes of the Wing's flying squadrons are on the rudder. (Don Logan)

F-15A 76-0010 seen at Luke AFB on February 28, 1988 is marked as the 405 Tactical Training Wing Flagship. (Douglas Slowiak/Vortex Photo Graphics)

Right: F-15A 76-0072 seen at Nellis AFB in August 1985 is marked as the 405 Tactical Training Wing Flagship. (Kevin Patrick)

F-15A 73-0107 seen here is wearing the tail stripe of the 550th TFS and marked as the 405 TTW Flagship. In addition the tail stripes of the Wing's flying squadrons are on the rudder. (Don Logan Collection)

F-15A 76-0046 seen at Luke AFB on September 8, 1990 is marked as the 405 Tactical Training Wing Flagship. (Kevin Patrick)

F-15E 87-0190 seen at Luke AFB on April 27, 1991 is marked as the 405 Tactical Training Wing Flagship. (Douglas Slowiak/Vortex Photo Graphics)

58th FIGHTER WING

461st TFTS, 550th TFTS, and 555th TFTS

In keeping with the new Air Force policy of one base – one Wing, the 405th TTW, training F-15E aircrews, was inactivated on October 1, 1991 and the squadrons, the 555th TFTS flying F-15A/B and the 461st and 550th TFTSs flying F-15E Strike Eagles, were transferred to the 58th FW, also based at Luke AFB. The aircraft were marked with the 58th FW's LF tail code. The other squadrons of the 58th FW, the 310th FS, the 311th FS, and the 314th FS were training F-16 pilots. The 550th FS was inactivated a month later on November 14, 1991. The 58th FW reassigned from TAC to the new ACC on June 1, 1992, followed on July 1, 1994 with reassignment from ACC to Air Education and Training Command (AETC). The 555th TFTS flying F-15A/Bs was inactivated on March 25, 1994, being activated on April 1, 1994 at Aviano AB, Italy as an F-16C squadron. As a result, the 550th was activated on March 25, 1994, taking over the resources of the 555th. The 58th FW was replaced by the 56th FW on April 1, 1994.

(Don Logan)

Left: F-15E 87-0185 seen at Luke AFB in May 1992 is marked as the 58th Fighter Wing Flagship. (Mick Roth)

Below: F-15E 87-0185 seen at Luke AFB on January 26, 1994 is marked as the 58th Fighter Wing Flagship. (Don Logan)

56th FIGHTER WING

461st FS, 550th FS, and 555th FS

In 1991, MacDill AFB was programmed for closure by Federal Base Closure and Realignment Commission. However, the 56th FW, then assigned to MacDill as an F-16 training unit, was identified as one of the Air Force's most illustrious units; so, after turning over operation of MacDill AFB to the 6th ABW in January 1994, the 56th FW moved without personnel or equipment to Luke AFB, Arizona. As a result of this move, the 58th FW was redesignated as the 56th FW on April 1, 1994. The 58th number was then assigned to the 542nd CTW at Kirkland AFB, New Mexico. At Luke, the 56th operated the formal training units for F-15E Strike Eagle crews and active-duty F-16 pilots. The 461st FS was inactivated on August 5, 1994, leaving the 550th FS as the single F-15E training squadron. The Air Force decided it was more economical to transfer the training function to the operational F-15E Wing at Seymour-Johnson AFB. The last F-15E class at Luke AFB graduated in February 1995. The last F-15E departed Luke on March 21, 1995 and the 550th FS inactivated on March 31, 1995, ending F-15 operations at Luke AFB, where the first production model F-15 had been delivered to the Air Force in November 1974, 21 years earlier.

Right: F-15E 87-0185 seen at Luke AFB in May 1992 is marked as the 56th Fighter Wing Flagship. (Kevin Patrick)

Below: F-15E 87-0185 seen at Luke AFB in June 1994 is marked as the 56th Fighter Wing Flagship. (Kevin Patrick)

426th TACTICAL FIGHTER TRAINING SQUADRON
KILLER CLAWS

The 426th TFTS was assigned to the 405th TTW as an F-15 squadron and received its first F-15 A/B aircraft on January 1, 1981. By January 1989 the 426th had received F-15D aircraft. 426th aircraft wore the LA tail code with red as their squadron color. As a result of the transfer of F-15 air superiority/interceptor training to Tyndall AFB, the 426th was inactivated on November 29, 1990.

Below: F-15A 76-0075 is seen here at Luke AFB in November 1983 marked as the 426th TFTS/AMU Flagship. (Kevin Patrick) Left: (Douglas Slowiak/Vortex Photo Graphics)

F-15A 73-0103 is seen here at Luke AFB in November 1983. Because the last three digits of the serial number could be repeated on more than one aircraft (example 73-0103 and 74-0103) non-standard serial number presentations using four large digits instead of three were used on some aircraft. (Kevin Patrick)

F-15A 76-0075 is seen here at Luke AFB on July 31, 1988. (Douglas Slowiak/Vortex Photo Graphics)

F-15D 82-0048 is seen here at Luke AFB on November 21, 1987. (Kevin Patrick)

461st FIGHTER SQUADRON
DEADLY JESTERS

The 461st TFTS was assigned to the 58th TTW as an F-15 squadron and received its first F-15 A/B aircraft on July 1, 1977. While assigned to the 405th TTW, the 461st received F-15D aircraft in January 1983. The 461st TFTS received their first F-15E on July 18, 1988, converting from air superiority variants. While assigned to the 58th TTW and 405th TTW, 461st aircraft wore the LA tail code with yellow as their squadron color. On October 1, 1991, the 461st was reassigned with the other F-15E squadrons to the 58th FW, with its F-15E aircraft changing their tail codes to LF. The 58th FW was redesignated as the 56th FW on April 1, 1994. The 461st FS was inactivated on August 5, 1994, as part of the relocation of F-15E aircrew training to the 4th FW at Seymour-Johnson AFB.

F-15A 77-0141 is seen here at Luke AFB on February 28, 1988 marked as the 461st TFTS Flagship. (Douglas Slowiak/Vortex Photo Graphics)

F-15E 86-0187 is seen here at Luke AFB on September 1, 1993 marked as the 461st FS Flagship. (Kevin Patrick)

F-15B 73-0109 is seen here at Luke AFB in September 1976 in Air Superiority Blue paint. (Mick Roth)

F-15B 73-0113 is seen here at Luke AFB on May 10, 1985. As with the 426th TFTS, As with other Luke aircraft, because the last three digits of the serial number could be repeated on more than one aircraft, the non-standard serial number presentations using four large digits instead of three were used on some aircraft. (Douglas Slowiak/Vortex Photo Graphics)

F-15A 75-0042 is seen here at Luke AFB on February 16, 1985. (Kevin Patrick)

F-15A 76-0083 is seen here at Luke AFB on July 31, 1988. (Jim Goodall)

F-15B 77-0163 is seen here at Luke AFB on March 13, 1986. (Douglas Slowiak/Vortex Photo Graphics)

F-15E 86-0186 is seen here at Luke AFB on July 31, 1988. (Douglas Slowiak/Vortex Photo Graphics)

F-15E 87-0169 is seen on July 31, 1988. (David F. Brown)

F-15E 87-0176 is seen here at Luke AFB on March 6, 1993. (Norris Graser)

F-15E 87-0205 is seen here at Luke AFB on March 17, 1993. (Kevin Patrick)

550th FIGHTER SQUADRON
SILVER EAGLES

The 550th TFTS was assigned to the 58th TTW as an F-15 squadron and received its first F-15 A/B aircraft on August 25, 1977. The 550th TFTS received their first F-15E on May 12, 1989, converting from air superiority variants. While assigned to the 58th TTW and 405th TTW, 550th aircraft wore the LA tail code with black and silver as their squadron colors. On October 1, 1991, the 550th was reassigned with the other F-15E squadrons to the 58th FW, with its F-15E aircraft changing their tail codes to LF. The 550th was inactivated a month later on November 14, 1991. With the inactivation of the 555th on March 25, 1994, the 550th was again activated taking over the resources of the 555th. As a result of the transfer of F-15E training to Seymour-Johnson AFB, the last F-15E class at Luke AFB graduated in February 1995, and the last F-15E departed Luke on March 21, 1995. The 550th FS inactivated on March 31, 1995, ending F-15 operations at Luke AFB.

F-15A 76-0078 is seen here at Luke AFB in August 1983 marked as the 550th TFTS/AMU Flagship. (Jim Goodall)

F-15E 87-0186 is seen here at Langley AFB on February 28, 1988 marked as the 550th FS Flagship. (Brian C. Rogers)

F-15B 73-0114 is seen here at Luke AFB on August 21, 1985. (Kevin Patrick)

F-15B 74-0137 is seen on June 20, 1983. (William R. Peake)

F-15A 76-0070 is seen in October 1987. (Jim Goodall)

F-15A 76-074 is seen here at Luke AFB in June 1985. (Jerry Geer)

F-15B 77-0155 is seen here at Luke AFB in June 1985. (Jerry Geer)

F-15D 78-0569 is seen here at Luke AFB on August 27, 1988. (Kevin Patrick)

Four F-15Es, two from the 461st FS and two from the 550th FS, are seen loaded with Mk 82 500 pound bombs inbound to the bombing range. (Boeing)

F-15E 87-0178 is seen here at Seymour-Johnson AFB in December 1990 marked as the 550th TFTS/AMU Flagship. The serial number presentation of 8-7187 on this aircraft is also non-standard. (David F. Brown)

F-15E 87-0190 is seen here at Luke AFB on May 20, 1994. (Norris Graser)

F-15E 87-0194 is seen here at Luke AFB on April 6, 1991. (Norris Graser)

F-15E 88-1679 is seen here at Luke AFB on May 20, 1994. (Norris Graser)

555th FIGHTER SQUADRON
TRIPLE NICKEL

The 555th TFTS was assigned to the 58th TFTW. The 555th TFTS Triple Nickel was the first operational squadron to receive an Eagle when it received a TF-15A (73-0108) christened TAC 1 after being accepted by President Gerald Ford on November 14, 1974. The 550th TFTS received its first F-15D in October, 1982. During October and November 1991, the 555th phased down its air-to-air F-15 operation, transferring all F-15A/B/C/D training to the 325th FW at Tyndall AFB. The Wing's last F-15A left on December 20, 1991. The 555th TFTS received its first F-15E in November, 1991. While assigned to the 58th TTW and 405th TTW, 555th aircraft wore the LA tail code with green as the squadron color. On October 1, 1991, the 555th was reassigned with the other F-15E squadrons to the 58th FW. While assigned to the 58th FW and the 56th FW, F-15Es wore LF tail codes. With the inactivation of the 550th TFTS a month later on November 14, 1991, the 555th took over the 550th's duties. The 555th was inactivated on March 25, 1994, being activated on April 1, 1994 at Aviano AB, Italy as an F-16C squadron.

Right: F-15A 76-0064 is seen here at Luke AFB in April 16, 1983 marked as the 550th TFTS/AMU Flagship. (Douglas Slowiak/Vortex Photo Graphics)

Below: F-15E 87-0190 is seen on August 23, 1992 marked as the 550th FS Flagship. (Norris Graser)

80-0062, a Royal Saudi Air Force F-15C in USAF markings is seen in a near vertical climb over the Arizona desert. (Boeing)

F-15A 76-0026 is seen in August 1991 marked as the 550th TFTS/AMU Flagship. (Jerry Geer)

F-15A 73-0085 is seen here at Luke AFB in December 1977 in Air Superiority Blue paint. (Don Logan Collection)

F-15A 73-0085 is seen here at Luke AFB on February 16, 1985. (Kevin Patrick)

F-15A 73-0090 is seen here at Luke AFB on July 1, 1995 in Air Superiority Blue paint. (Dennis R. Jenkins)

F-15A 73-0091 is seen here at Luke AFB in June 1975 in Air Superiority Blue paint. (Don Logan Collection)

F-15B 73-0110 is seen here at Luke AFB on March 13, 1986 in Air Superiority Blue paint. (Gary Emery)

F-15A 76-0070 is seen on June 4, 1991. (Norris Graser)

F-15D 79-0014 is seen here at Luke AFB on February 16, 1985. (Kevin Patrick)

F-15D 81-0063 is also seen at Luke AFB on February 16, 1985. (Kevin Patrick)

F-15E 87-0187 is seen here at Luke AFB in May 1992. (Mick Roth)

F-15F 87-0189 (above) and F-15E 89-0477 (Below) are seen here at Luke AFB in August 1993. (Both Jerry Geer)

MOUNTAIN HOME AFB, IDAHO, 366th WING

390th FS F-15C/D and 391st FS F-15E

In early 1991, the Air Force announced that the 366th Fighter Wing would become the Air Force's Air Intervention Composite Wing. The dynamic, five squadron wing has the ability to deploy rapidly and deliver integrated combat airpower. The Air Intervention Composite Wing's rapid transition from concept to reality began in October 1991 when Air Force redesignated the 366th Tactical Fighter Wing as the 366th Wing. The wing's newly reactivated fighter squadrons became part of the Composite Wing in March 1992. The 389th Fighter Squadron began flying the dual-role F-16C Fighting Falcon, while the 391st Fighter Squadron was equipped with the new F-15E Strike Eagle. These two squadrons provide the Wing round-the clock precision strike capability. Like all 366th Wing aircraft the F-15s wear MO (Mountain Home AFB) tail codes and tail stripes corresponding to the squadron color.

In June 1992, as part of Air Force restructuring, Air Combat Command replaced Tactical Air Command. A month later, the 366th also gained the 34th Bomb Squadron. The 34th was located at Castle AFB, California, and flew the B-52G Stratofortress, giving the composite wing deep interdiction bombing capabilities. In 1994, after the B-52G retired, the 34th Bomb Squadron picked up B-1Bs. In September 1992, Air Force redesignated the 390th Electronic Combat Squadron as the 390th Fighter Squadron, which began flying the F-15C Eagle. Also during October 1992, the Composite Wing gained its final flying squadron when the 22nd Air Refueling Squadron was activated and equipped with the KC-135R Stratotankers.

During 1993, and again in 1995, the 366th served as the lead unit for Operation Bright Star, a large combined exercise held in Egypt. In July 1995, the Wing verified its combat capability in the largest operational readiness inspection in Air Force history when it deployed a composite strike force to Cold Bay, Canada, and proved they could deliver effective composite airpower. In 1996, the Wing also deployed a composite force in support of Operation Provide Comfort in Turkey, and in 1997 launched another composite force to support Operation Southern Watch in Saudi Arabia.

(Don Logan)

Below: F-15C 86-0148 is seen on a stop over at Tinker AFB on October 22, 1993 marked as the 366th Wing Flagship. (Don Logan)

158

F-15C 86-0150 is seen taxiing at Nellis AFB in October 1995. It is marked as the 366th Operations Group Flagship. (David F. Brown)

F-15E 87-0182 is seen on a stop over at Langley AFB on October 11, 1994 marked as the 366th Wing Flagship. (Brian C. Rogers)

390th FIGHTER SQUADRON
WILD BOARS

The 390th Fighter Squadron began as a fighter squadron in World War II. The unit activated on June 1, 1943 at Richmond, Virginia as part of the 366th Fighter Group. The 390th TFS flew F-4D Phantom II from Da Nang AB, Republic of South Vietnam beginning in 1969 until the unit moved to Mountain Home AFB on June 30, 1972. At Mountain Home AFB, the 390th began flying F-111Fs. In 1977, the unit still based at Mountain Home AFB, converted to the F-111A. On December 15, 1982 the squadron was redesignated the 390th Electronic Combat Squadron as it converted to the EF-111. The 390th ECS deployed EF-111s in support of Operation Desert Storm, jamming Iraqi radar and communication systems. With the transfer of the EF-111s to Cannon AFB, New Mexico, the squadron inactivated on September 11, 1992 and reactivated on September 25, 1992, as the 390th Fighter Squadron flying air superiority F-15C/D's. Since the squadron changed to the F-15C, the squadron has participated in Operations Provide Comfort and Southern Watch.

Below: F-15C 86-0151 is seen here at Nellis AFB on April 19, 1994 marked as the 390th Fighter Squadron Flagship. (Craig Kaston)

F-15C 86-0146 is seen taxiing at Volk Field, Wisconsin on June 11, 1993. (Norris Graser)

F-15D 86-0181 is seen in September 1992 with the low visibility tail markings. (Ben Knowles)

F-15C 86-0162 is seen on a stop over at Tinker AFB in June 1994. (Jim Geer)

F-15D 86-0181 is seen here at Volk Field, Wisconsin on June 11, 1993. (Norris Graser)

391st FIGHTER SQUADRON
BOLD TIGERS

The growing involvement of the United States military in World War II caused the creation of the 391st Fighter Squadron. On 24 May 1943 the Army constituted the 391st Fighter Squadron as part of the 366th Fighter Group. Squadron operations throughout the 1970s and 1980s centered on maintaining combat capability and training F-111 aircrew. The 391st Tactical Fighter Squadron ended its F-111A mission on July 1, 1990, and the squadron was inactivated when the retiring of the F-111A fleet.

With the 366th Tactical Fighter Wing redesignated as the 366th Wing in preparation for a change to a composite force structure, the 391st once again became activated as a component of the Wing. When activated on March 11, 1992, the squadron was redesignated as the 391st Fighter Squadron and equipped with the F-15E Strike Eagle, serving as a dual-role asset to the Air Force's first air intervention composite Wing.

(Boeing

F-15E 87-0210 is seen taxiing at McChord AFB, Washington in December 1994. It is marked as the 391st Fighter Squadron Flagship. (R.E.F. Jones)

F-15E 89-0497 is seen here at Nellis AFB on April 19, 1994. The aircraft crashed near McDermott, Oregon on October 20, 1998 during a night training mission. Both the Pilot and WSO were killed. (Craig Kaston)

F-15E 88-1705 is seen here at Volk Field, Wisconsin on June 6, 1993. (Norris Graser)

F-15E 87-0202 is seen taxiing at Mountain Home AFB in September 1992 carrying CBUs. (Ben Knowles)

F-15E 87-0209 is seen here at Volk Field, Wisconsin on June 6, 1993. (Nate Leong)

F-15E 87-0170 is seen here at Mountain Home AFB in August 1997. (Jim Geer)

NELLIS AFB, NEVADA

HEADQUARTERS AIR WARFARE CENTER
53rd Wing, Eglin Air Force Base, Florida
57th Wing, Nellis Air Force Base, Nevada
99th Air Base Wing, Nellis Air Force Base, Nevada

USAF WEAPONS SCHOOL
F-15 Division
F-15E Division

AIR WARFARE CENTER

The Air Warfare Center, established in October 1995 and located at Nellis Air Force Base, Nevada, manages advanced aircrew training and integrates many of the Air Force's test and evaluation requirements. It is an outgrowth of the USAF Tactical Fighter Weapons Center, established in 1966, which concentrated on the development of forces and weapons systems that were specifically designed for tactical air operations in conventional (non-nuclear) war. It continued to perform this mission for nearly thirty years, undergoing several name changes in the 1990s. In 1991, the center became the USAF Fighter Weapons Center, and then the USAF Weapons and Tactics Center in 1992, and finally the Air Warfare Center.

Throughout the history of the Center, the 57th Wing has been the Center's flying unit.

The Air Warfare Center uses the Nellis Air Force Range Complex, occupying about three million acres of land, the largest such range in the United States, and another five-million-acre military operating area which is shared with civilian aircraft. The Center also uses the range complex at Eglin AFB, Florida that adds even greater depth to the Center's capabilities, providing over water and additional electronic expertise to the Center. The Air Warfare Center oversees operations of the 57th Wing and 99th Air Base Wing at Nellis AFB and the 53rd Wing at Eglin AFB, Florida.

F-15C 84-0030 assigned to the 422nd TES is seen on a training mission over Lake Mead near Las Vegas, Nevada. (USAF) Place between title "AIR WARFARE CENTER" and text.

FIGHTER WEAPONS CENTER (TFWC and FWC)

Above: F-15D 81-0063 is seen in June 1992 marked as the Fighter Weapons Center Flagship. (David F. Brown)

Right: F-15D 82-0044 is seen here at Nellis AFB in June 1990 marked as the Tactical Fighter Weapons Center Flagship. (Ben Knowles)

Below: F-15C 82-0013 is seen here at Nellis AFB in June 1990 marked as the Tactical Fighter Weapons Center Flagship. (Douglas Slowiak/Vortex Photo Graphics)

WEAPONS AND TACTICS CENTER (WTC)

F-15D 81-0065 is seen here at Tinker AFB on October 9, 1992 marked as the Weapons and Tactics Center (WTC) Flagship. (Jerry Geer)

F-15D 82-0044 is seen here at Langley AFB on July 7, 1993 marked as the Weapons and Tactics Center (WTC) Flagship. (Brian C. Rogers)

57th WING

The 57th Wing started out as the 57th Fighter Weapons Wing (FWW), which replaced the 4525th FWW on October 15, 1969. On April 1, 1977 the 57th was redesignated 57th Tactical Training Wing (TTW). It was once again designated the 57th FWW on March 1, 1980, then the 57th Fighter Wing (FW) on October 1, 1991, and given its current designation as the 57th Wing on June 15, 1993.

The 57th FWW was one of the first users of the F-15 as TAC's test and evaluation unit for fighter weapons and tactics. The 57th has traditionally used a yellow and black checked tail stripe and WA tail codes, reportedly standing for Weapons Acquisition. The Wing received its first F-15, and F-15A on October 1, 1976, after operating several of the original test aircraft as Detachment 1 from Luke AFB for short periods during the summer of 1976. In these early tests the aircraft wore Luke AFB's LA tail code but had the Wing patch and tail stripe of the 57th FWW.

The 57th Wing is responsible for a variety of activities at Nellis, such as Red Flag, which provides realistic training in a combined air, ground, and electronic threat environment for U.S. and allied forces; Air Force graduate-level weapons and tactics training (Weapons School) for A-10, B-1, B-52, EC-130, E-3, F-15, F-15E, F-16, HH-60, RC-135, command and control operations, intelligence, and weapons instructor academic and flying courses; plans and executes close air support missions, such as Air Warrior, in support of U.S. Army exercises and interoperability training with the Army; the Air Force's air demonstration team the Thunderbirds; and the operation and deployment of the Predator, an unmanned reconnaissance aircraft. The Wing also flies HH-60G PAVEHAWK helicopters in support of combat rescue as well as rescue in the Nellis Air Force Range Complex and nearby civilian communities.

Below: (USAF)

F-15C 82-0026 is seen here at Langley AFB on October 6, 1989 marked as the 57th Fighter Weapons Wing Flagship. (Brian C. Rogers)

F-15C 82-0026 is seen on October 19, 1989 marked as the 57th Fighter Weapons Wing Flagship. The Eagle applied to the inside of both vertical stabilizers is visible on the left vertical stabilizer. (Phillip Huston)

F-15C 82-0031 is seen here at Nellis AFB on October 6, 1989 marked as the 57th Fighter Weapons Wing Flagship. (Craig Kaston)

F-15C 82-0026 is seen here at Nellis AFB on October 6, 1989 marked as the 57th Wing Flagship. (Brian C. Rogers)

F-15E 86-0190 is seen here at Langley AFB on October 6, 1989 marked as the 57th Test Group Flagship. (David F. Brown)

F-15A 75-0042 is seen here at Nellis AFB on December 3, 1976 carrying the early white tail codes and serial number presentation. (Don Logan)

F-15A 75-0055 is seen here at Nellis AFB on March 31, 1978 carrying the early white tail codes and serial number presentation. (Don Logan)

F-15B 77-0162 is seen here at Offutt AFB in February 1992. In this earlier version, a light gray Eagle contained in a black stripe is applied to the inside of both vertical stabilizers. (George Cockle)

F-15A 74-0124 is seen here at Nellis AFB in January 1979 carrying black tail codes and serial number presentation. (David F. Brown)

A two ship formation of 57th FWW F-15As 76-0119 and 76-0120 banks over the Nevada desert north of Nellis AFB. (USAF)

F-15B 76-0141 is seen here at Nellis AFB on July 16, 1979. Overspray from the recently applied black tail codes and serial number presentation is visible on the vertical stabilizer. (Brian C. Rogers)

F-15C 82-0022 is seen here at Luke AFB on December 21, 1985, with the light gray Eagle in a black stripe applied to the inside of both vertical stabilizers. (Douglas Slowiak/Vortex Photo Graphics)

F-15C 82-0025 is seen taxiing from its parking space on the ramp at Nellis AFB on June 23, 1990. (Bob Greby)

F-15C 82-0031 is seen here at Nellis AFB on April 12, 1985. (Douglas Slowiak/Vortex Photo Graphics)

F-15C 85-0121 is seen here at Nellis AFB in June 1990. (Don Logan)

F-15E 86-0190 is seen in this shot test firing an AIM-120 AMRAAM. (USAF)

F-15E 86-0190 is seen in December 1993 marked as the 57th Test Group Flagship. (Jim Geer)

STRIKE EAGLE CAMOUFLAGE TEST SCHEMES

Three F-15Cs assigned to the 57th at Nellis were painted in different schemes to determine the production F-15E camouflage. 82-0022 was painted in the very dark gunship gray (FS36118). 82-0028 used three tones of gray in a pattern similar to one used on one of the F-16XL SCAMP prototypes. 82-0029 was painted in a Compass Ghost pattern using a lighter and darker shade of gray than usual (FS36251 and FS36176). The dark gunship gray was the scheme chosen for use by production F-15Es.

Three F-15Cs, 82-0022, 82-0028, and 82-0029, while assigned to the 422nd TES, were used as test aircraft for F-15E Strike Eagle paint schemes. 82-0022 is seen and in the picture below in the single color gunship gray scheme later adopted for use on the F-15E. (Both Craig Kaston)

F-15C 82-0028 was painted with a three-tone scheme similar to the one used on the F-16XL. (Both Craig Kaston)

F-15C 82-0029 was painted with a two tone gray scheme later modified into the Mod Eagle scheme used on most air superiority Eagles. (Both Craig Kaston)

57th Fighter Weapons Wing – USAF Tacticval Fighter Weapons Center - Detachment I

57th Fighter Weapons Wing – Detachment I was responsible for Operational Test and Evaluation of F-15 weapons. Testing was accomplished at both Nellis and Luke AFB. The aircraft were based at Luke AFB, and wore LA tail codes, but carried 57th Fighter Weapons Wing markings including the yellow and black checkerboard tail stripe.

F-15B 73-0108 (TAC-1) was used for operational test and evaluation early in the Eagle's career. At the time of this testing all of TAC's Eagles were stationed at Luke AFB. The aircraft deployed to Nellis for testing which required the use of the Nellis Range Complex. TAC-1 is seen taxiing for a test mission at Nellis AFB on July 30, 1975. (Don Logan)

F-15B 73-0108 is seen here at Nellis on July 22, 1975. (Don Logan)

TAC-1 is seen on the ramp at Luke AFB in August 1975. (Jim Rotramel)

Right: F-15A 73-0089 is seen here at rotation on takeoff at Nellis AFB on August 12, 1975. (Don Logan)

F-15A 73-0087 is seen taxiing back from a test mission at Nellis AFB on July 22, 1975. (Don Logan)

F-15A 73-0089 is seen on the ramp at Nellis AFB on July 22, 1975. (Don Logan)

F-15A 73-0085 is seen here at on the ramp at Nellis AFB on July 22, 1975. (Don Logan)

F-15A 73-0102 is seen here at on the ramp at Nellis AFB during follow-on weapons testing on November 18, 1975. (Don Logan)

F-15A 73-0105 is seen here at Nellis AFB on November 18, 1975. It is carrying a GBU-8 Electro-Optical Guided Bomb (EOGOB) TV guided 2000-pound bomb. This early weapon differed from the GBU-15, which replaced it in that the GBU-8 also locked onto the target before release, but could not have the target updated in flight. (Don Logan)

422nd TEST AND EVALUATION SQUADRON
BATS

The 422nd Test Evaluation Squadron (TES) was originally desig-nated the 422nd Fighter Weapons Squadron (FWS). The squadron provides the continual refinement of combat techniques and tac-tics, including weapon release profiles, maneuvering, and electronic warfare tactics. The 422nd received F-15A/Bs in 1982, followed by F-15C/Ds, and in 1990, F-15Es. It was assigned to the 57th Wing from October 15, 1969 until transferring to the 53rd Wing at Eglin AFB on October 1, 1996. Presently, the 422nd TES is a composite squadron assigned to the 53rd Wing at Eglin AFB Florida, and oper-ates OT tail coded aircraft from Nellis AFB. It conducts opera-tional tests of A-10, F-15C, F-15E, F-16C and HH-60G hardware and software enhancements prior to release to the Combat Air Forces. The squadron develops and evaluates tactics to optimize the combat capability of these weapon systems in a simulated com-bat environment. The 422nd TES also develops and publishes new tactics for these aircraft. The results of these tests directly benefit aircrews in ACC, PACAF, and USAFE by providing them with op-erationally proven hardware and software systems.

Two above: (Douglas Slowiak/Vortex Photo Graphics)

F-15C 84-0012 is seen here at Nellis AFB marked as the 422nd Test and Evaluation Squadron (TES) Flagship. (Douglas Slowiak/Vortex Photo Graphics)

Right: This rear photo of F-15C 84-0012 Nellis AFB shows the 422nd Test and Evaluation Squadron (TES) Flagship markings. (Douglas Slowiak/Vortex Photo Graphics)

Below: F-15E 86-0190 marked as the 422nd TES Flagship is loaded with MK 20 Rockeye Cluster Bombs. (Boeing)

F-15E 86-0190 is seen here at Luke AFB marked as the 422nd Test and Evaluation Squadron (TES) Flagship. (Ben Knowles)

F-15C 83-0037 is seen here at Nellis AFB on September 29, 1997 wearing the OT tail code of the 422nd TES, 53rd Wing based at Nellis AFB. (Don Logan)

F-15E 92-0365 is seen here at Nellis AFB on September 29, 1997 also wearing the OT tail code of the 422nd TES, 53rd Wing. (Don Logan)

F-15C 81-0030 is also seen at Nellis AFB on September 29, 1997. (Don Logan)

USAF WEAPONS SCHOOL

The USAF Weapons School, as it is known today, developed from a corps of World War II fighter pilots highly skilled in aerial combat. From that informal beginning, the USAF Fighter Gunnery School was established. Because of heavy commitments to the Korean War effort, the USAF Fighter Gunnery School was converted to combat crew training. This mission would last through the end of 1953. Effective January 1, 1954, the school graduated its last Combat Crew Training Class and assumed, as its primary mission, the training of gunnery instructors for the USAF. On that date, the squadron received a new title-the USAF Fighter Weapons School. During the fifties, the F-51, F-80, F-84, and F-100A were the primary aircraft used for instruction at the school. Soon after, the F-100C and F-100D entered the school inventory, providing complete mission capability.

Mid-1965 marked the beginning of another expansion of the Fighter Weapons School, as F-4 and F-105 Fighter Weapons Instructor Courses were established. Due to the expanding scope of the school, it was converted into the 4525th Fighter Weapons Wing in 1966 and was comprised of three squadrons; an F-100 squadron, an F-4 squadron, and an F-105 squadron. In 1969, the 4525th became the 57th Fighter Weapons Wing (FWW). During 1992, as a result of the creation of ACC, bomber aircraft from the disbanded SAC were added and the B-52 and B-1 divisions were formed from SAC's strategic bomber weapons school. As a result of the addition of bomber aircraft "Fighter" was dropped from the school name leaving it the USAF Weapons School. The school is now comprised of Divisions for the A-10, B-1, B-52, EC-130, E-3, F-15C, F-15E, F-16C, HH-60, and RC-135. Each Division operates as an advanced 'college' where tactics and weapons employments are taught to a cadre of experienced flight crews. These crews then return to their operational units to pass on the information. Both the F-15 Division and the F-15E Division are located at Nellis AFB.

(Don Logan)

183

F-15C 82-0027 is seen on landing approach to Nellis. It is marked as the Fighter Weapons School (FWS) Flagship. (Don Logan Collection)

F-15C 80-0033 is seen here at Nellis AFB on September 29, 1997 marked as the WEAPONS SCHOOL commander's aircraft. (Don Logan)

F-15E 90-0227 is also seen at Nellis AFB on September 29, 1997 marked as the WEAPONS SCHOOL commander's aircraft. (Don Logan)

433rd FIGHTER WEAPONS SQUADRON
SATIN'S ANGELS

In October 1976, after arrival of the 57th FWW's first F-15, the 433rd Fighter Weapons Squadron (FWS) began developing the syllabus for the F-15 Fighter Weapons Instructor Course. On January 3, 1978 it received its first Fighter Weapons Instructor class. For most of the next year, through the beginning of 1979, the 433rd conducted F-15 pilot training (RTU) classes. The 433rd was devoted exclusively to conducting the F-15 Fighter Weapons Instructor Course from March 1, 1979 until consolidated into the Fighter Weapons School as the F-15 Division on June 1, 1981. The 433rd was inactivated on December 30, 1981. It was later activated at Holloman AFB as part of the 479th TTW flying AT-38s.

F-15A 77-0141 is seen here at Nellis AFB on December 8, 1980 carrying the devil head of the 433rd FWS on the inside of the vertical stabilizer. (Brian C. Rogers)

F-15B 77-0164 is seen here at Nellis AFB also on December 8, 1980 carrying the same 433rd FWS devil head as the single seat A models. (Brian C. Rogers)

F-15 DIVISION

The F-15 Division of the Fighter Weapons School was established on December 30, 1981 taking over F-15 Weapons Instructor training from the 433rd FWS. The course is focused on air-to air combat and is made up of over 250 hours of classroom instruction and 36 flight hours.

Below: F-15C 80-0033 is seen here at Nellis AFB on September 29, 1997 marked as the WEAPONS SCHOOL F-15 Division Commander's aircraft. (Don Logan)

F-15C 80-0019 is seen here at Nellis AFB on September 29, 1997 with the Weapons School F-15 Division emblem on the left intake. (Don Logan)

F-15C 80-0049 is seen taxiing at Nellis AFB in November 1993. (David F. Brown)

F-15C 80-0043 is seen here at Nellis AFB on September 29, 1997 with the darker radome from an F-15E installed. (Don Logan)

F-15D 81-0062 is seen here at Nellis AFB on September 29, 1997 with the Weapons School F-15 Division emblem on the left intake. (Don Logan)

F-15E DIVISION

The F-15E Division of the Fighter Weapons School was established on July 8, 1991. The course is focused on both air-to-ground combat and air-to air combat and is made up of over 310 hours of classroom instruction and 50 flight hours.

(Don Logan)

F-15E 90-0227 is seen marked as the WEAPONS SCHOOL F-15E Division Commander's aircraft. The F-15E weapons school emblem is visible on the left intake. (David F. Brown)

F-15E 88-1677 is seen on the Nellis AFB ramp marked as the WEAPONS SCHOOL F-15E Division Commander's aircraft. (Alec Fushi)

F-15E 88-1681 is seen in July 1990 with the Weapons School emblem visible on the right intake. (Ben Knowles)

F-15E 89-0481 is seen in June 1993. (David F. Brown)

F-15E 86-0189 is seen in June 1990. (Ben Knowles)

F-15E 89-0475 is seen in June 1995. (David F. Brown)

F-15E 90-0228 is seen taxiing at Nellis AFB in March 1996. (Alec Fushi)

SEYMOUR-JOHNSON AFB, NORTH CAROLINA, 4th FIGHTER WING

333rd, 334th, 335th, 336th Fighter Squadrons all flying the F-15E

The 4th Fighter Wing, stationed at Seymour-Johnson AFB, North Carolina is the only all F-15E wing in the USAF. The wing is made up of four fighter squadrons, the 333rd, 334th, 335th, and 336th Fighter Squadrons. All of the Fighter Wing's aircraft carry a SJ tail code for Seymour-Johnson AFB.

The 4th Wing is one of the most distinguished fighter units in the world. It has the distinction of being one of only two Air Force units that can trace its history to another country. Before the United States' entry into World War II, American volunteers were already flying combat in Royal Air Force Eagle Squadrons (71st, 121st, and 133rd). When the United States entered the war, these units, and the American pilots in them, were transferred to the U.S. Army Air Forces, 8th Air Force, forming the 4th Fighter Group on September 12, 1942. In World War II the 4th Fighter Group (with squadrons now designated the 334th, 335th, and 336th) flew missions during the air war over Europe. The group was credited with the destruction of 1,016 enemy aircraft, more than any other comparable 8th Air Force unit, and produced 37 aces. The 4th Fighter Group was inactivated at Camp Kilmer, New Jersey in November 1945.

The unit was reactivated at Selfridge Field, Michigan, almost a year later flying F-80 Shooting Star aircraft. In March 1948, the 4th Fighter Group (FG) transitioned to F-86 Sabre jets. In December 1950, the group (now designated the 4th Fighter-Interceptor Group) was the first unit to commit F-86 Sabre jets in the Korean War, shooting down a MiG-15 December 17 during the first Sabre jet mission of the war. Four days later elements of the 4th FG were involved in the first all-jet fighter battle in history. The flight element downed six MiG-15s without sustaining any losses. While in Korea, the unit was redesignated as the 4th Fighter Interceptor Wing (FIW). During the Korean War pilots of the 4th destroyed 502 enemy aircraft (54 percent of the total), becoming the top fighter unit of the Korean conflict. Twenty-four pilots achieved ace status.

The group moved to Japan following the Korean armistice in 1953, continuing training and tours to Korea. The unit moved to Seymour-Johnson Air Force Base, North Carolina, December 8, 1957, picking up a fourth squadron, the 333rd Lancers. The 4th FG was redesignated as the 4th Tactical Fighter Wing in July 1958. Aircrews of the 4th Fighter Wing flew F-100 Super Sabre aircraft at the new location and, within two years, the Wing became the first Air Force unit to convert to F-105 Thunderchief aircraft. The 4th TFW transitioned to F-4D Phantom II aircraft beginning in early 1967. Flying F-4Es, the 4th TFW deployed elements to Southeast Asia beginning in April 1972. Operating from Ubon Royal Thai Air Base, Thailand, as the first F-4 wing to augment elements of Pacific Air Forces, aircrews of the 4th flew more than 8,000 combat missions, many into the heart of North Vietnam. A fourth F-4E Squadron, the 337th Falcons were assigned to the 4th TFW from July 1, 1982 until September 30, 1985.

(Don Logan)

The unit transitioned from the F-4E to the F-15E beginning on December 29, 1988 when the Wing's first F-15E, the "Spirit of Goldsboro" arrived. The 336th Tactical Fighter Squadron became the first operational F-15E squadron on October 1, 1989. At the height of conversion training, the 4th Tactical Fighter Wing was one of the first units tasked to react to Iraq's invasion of Kuwait Aug. 2, 1990. Two F-15E tactical fighter squadrons were deployed to Southwest Asia in August and December 1990, and led nighttime strikes against Iraqi forces on January 16, 1991, helping to bring the Persian Gulf War to a successful conclusion on February 28.

The transition from F-4E to the F-15E was completed on July 1, 1991, making the 4th the first operational F-15E wing in the Air Force. On April 22, 1991, the 4th Tactical Fighter Wing was redesignated the 4th Fighter Wing and incorporated the assets of the 68th Air Refueling Wing, a former Strategic Air Command unit flying the KC-10 tanker. With the addition of the tankers, the unit dropped "Fighter" from its designation becoming the 4th Wing. The 4th Wing became the 4th Fighter Wing again as a result of another force structure change, which occurred in 1994. All KC-10s were reassigned to the Air Mobility Command, and the 333rd FS returned to Seymour-Johnson as the F-15E Formal Training Unit (FTU), a function previously performed by the 550th FS of the 56th FW at Luke. The wing uses SJ tail codes (Seymour-Johnson AFB).

Three Flagships F-15E, the 4th Wing, 336th TFS and 335th TFS, are seen at "last chance" prior to taking the runway for takeoff. (Jeff Wilson)

F-15E 89-0472 is seen here at Seymour-Johnson AFB on October 16, 1998 marked as the 4th FW Flagship. A 335th Fighter Squadron emblem is visible on the left intake. (Don Logan)

F-15E 89-0494 is seen here at Seymour-Johnson AFB in May 1997 marked as the 4th FW Flagship. A 336th Fighter Squadron emblem is visible on the left intake. (Alec Fushi)

Above: F-15E 89-0504 is seen here at Seymour-Johnson AFB in July 1991 marked as the 4th WG (Wing) Flagship. A 336th Fighter Squadron emblem is visible on the left intake. (David F. Brown)

Right: F-15E 89-0472 is seen here at Seymour-Johnson AFB on October 16, 1997 marked as the 4th Operations Group (OG) Flagship. A 335th Fighter Squadron emblem is visible on the left intake. (Alec Fushi)

333rd FIGHTER SQUADRON
LANCERS

The 333rd Fighter Squadron was activated as the 333rd Fighter-Day Squadron on December 8, 1957, at Seymour-Johnson Air Force Base, North Carolina. It was assigned to the 4th Fighter-Day Wing and equipped with the F-100 Super Sabre. Redesignated the 333rd Tactical Fighter Squadron July 1, 1958, the squadron was assigned to the 355th Tactical Fighter Wing, Takhli Royal Thai Air Base, Thailand, on Dec. 8, 1965. Flying the F-105 Thunderchief, 333rd aircrews flew strike, air-to-air, armed reconnaissance, and close air support combat missions during the Vietnam War. The squadron was credited with 6.5 MiG-17 aerial victories, the most by a single squadron within the 355th Wing.

In October 1970, the 333rd was assigned to the 23rd Tactical Fighter Wing at McConnell Air Force Base, Kansas. The unit was redesignated the 333rd Tactical Fighter Training Squadron March 22, 1971, and assigned to the 58th Tactical Fighter Training Wing at Luke Air Force Base, Arizona. On July 1, 1971, the squadron moved to Davis-Monthan Air Force Base, Arizona, and by the end of the month it was once again under the 355th Tactical Fighter Wing. The unit's mission was to train student pilots in flying the A-7D Corsair II. Six years later, the squadron transitioned to the A-10A Thunderbolt II. The squadron was inactivated February 15, 1991. It was redesignated and activated as the 333rd Fighter Squadron on November 1, 1991, and assigned to the 602nd Air Control Wing. On May 1, 1992, it was reassigned to the 355th Operations Group as the first formal OA-10 training squadron.

The 333rd Fighter Squadron was moved to Seymour-Johnson from Davis-Monthan AFB, Arizona, without personnel or equipment, effective October 1, 1994. Under the 4th Fighter Wing's, 4th Operations Group, the squadron's new mission was to conduct formal training for F-15E aircrew members. Training consists of four courses: basic, transition, instructors, and senior officer checkout. The basic course lasts seven months and trains new pilots and weapon systems officers to fly the F-15E. The four month transition course is for experienced fighter crews who are changing to the F-15E from another first line fighter type. The instructors' course lasts two and one half months and involves training experienced F-15E aircrews how to become instructors. The senior officer checkout is a one month course intended to familiarize senior officers who have been assigned to a base with F-15Es.

(Alec Fushi)

Above and below: F-15E 89-0473 is seen in these two photographs, taken in October 1994, marked as the 333rd FS Flagship. GBU-12 500-pound class laser guided bombs are visible on the CFTs. (Both AF)

Right: F-15E 86-0186 is seen here at Seymour-Johnson AFB on October 16, 1998 in 333rd FS markings. (Don Logan)

F-15E 87-0185 is seen here at Seymour-Johnson AFB on October 16, 1998 taxiing for a training mission. (Don Logan)

F-15E 88-1684 is seen here at Seymour-Johnson AFB in September 1997. (Alec Fushi)

F-15E 89-0489 is seen here at Seymour-Johnson AFB on October 16, 1998 taxiing for a training mission. (Don Logan)

334th FIGHTER SQUADRON
EAGLES

The 334th was constituted by War Department letter on Aug. 22, 1942, and activated at Bushey Hall, England, on September 12, along with its sister squadrons, the 335th and 336th. The three Eagle squadrons, formerly composed of American volunteers in the Royal Air Force, were assigned to the 4th Fighter Group – the first Army Air Corps unit activated in the European Theater during World War II. As former members of RAF 71st Squadron, the 334th fighting Eagles continued to fly British Spitfires until the arrival of the P-47 Thunderbolt aircraft in 1943. About a year later the squadron changed to the P-51 Mustang, which served as the primary aircraft for the remainder of the war.

The 334th, with its sister squadrons took part in air battle accomplishments against the German Luftwaffe. The 334th scored 395 kills- 210 in the air and 185 on the ground. As a part of the Korean conflict, the 334th, equipped with F-86 Sabre Jets deployed to Korea and was credited with the destruction of 142.5 enemy aircraft during the Korean Conflict. The 334th returned to December 8, 1957, when it was reassigned to Seymour-Johnson AFB as a unit of the 4th Fighter Day Wing. The unit flew the F-100 Super Sabres until mid-1959 when transition to the F-105 Thunderchief aircraft began. After a six-month tour of Southeast Asia, the 334th returned to Seymour-Johnson in February 1966 and began instructing new pilots in F-105 operations. The Fighting Eagles rushed to Korea in January 1968, supporting operations during the Pueblo incident. The squadron returned to Seymour-Johnson in June 1968.

The 334th flew its first F-15E sorties on January 1, 1991. During this month the squadron served as the host unit for several units deploying to Operation Desert Shield and Desert Storm. Additionally, 334th aircrews and support personnel deployed to Operation Desert Storm as augmentees. The squadron became operational in the F-15E on June 18, 1991, and deployed to Saudi Arabia the next day to relieve remaining elements of the 335th, providing combat air patrol and ground alert forces supporting withdrawal of troops from Operation Desert Storm.

F-15E 87-0200 is seen in this photograph, taken on October 16, 1998, marked as the 334th FS Flagship. SUU-64/B Tactical Munitions Dispensers are visible on the CFTs. (Don Logan)

F-15E 89-0491 is seen in this photograph, taken in April 1991, marked as the 334th TFS Flagship. (David F. Brown)

F-15E 89-0491 is seen in this photograph, taken in October 1994, marked as the 334th FS Flagship. This aircraft crashed and was destroyed on July 11, 1997 as a result of an engine fire. (Alec Fushi)

F-15E 87-0193 is seen here at Seymour-Johnson AFB on October 16, 1998 preparing to taxi on a training mission. (Don Logan)

F-15E 88-1700 is seen here at stop over at Luke AFB in December 1993. (Douglas Slowiak/Vortex Photo Graphics)

F-15E 89-0500 is seen here at Seymour-Johnson AFB in October 1997. (Alec Fushi)

335th FIGHTER SQUADRON
CHIEFS

The 335th Fighter Squadron traces its ancestry back to the Royal Air Force 121 Squadron, formed on May 14, 1941 as the second of three "Eagle Squadrons" of the RAF. The "Chief's Head" insignia dates back to the original emblem of the RAF 121 Squadron, which featured in its center a profile of the head of an American Indian Chief. Following the entry of the US into the war, the three Eagle Squadrons were transferred to the US Army Air Forces as the 334th, 335th, and 336th Fighter Squadrons, then combined to form the 4th Fighter Group flying P-47 Thunderbolts and P-51 Mustangs. During World War II, the 335th Fighter Squadron destroyed 262 enemy aircraft. The Chiefs flew P-80 Shooting Stars until 1949 when they received the F-86 Sabre Jet, which they took to Korea on November 10, 1950. By the end of the Korean War the 335 Fighter Squadron led all squadrons with 218.5 kills. The Chiefs remained in the Far East until December 8, 1957, when they moved to Seymour-Johnson AFB, North Carolina and converted to the F-100 Super Sabre. In May of 1958, the 335th Fighter Squadron became the first squadron in the Air Force to receive the F-105 Thunderchief. They were sent to Eglin AFB Florida, and tasked with operational testing of the new aircraft for the next three years.

In 1967, the Chiefs received the airframe they would fly for the next twenty-three years – the F-4 Phantom II. On March 1, 1990, in conjunction with the fifty-first change of command, the last F-4 sortie and first F-15E sortie were flown. The 335th TFS achieved IOC on October 1, 1990, and on December 27-28, 1990, the 335th Fighter Squadron deployed twenty-four F-15Es along with support personnel and equipment to Al Kharj Air Base in central Saudi Arabia. On the night of January 17, the Chiefs participated in the initial assault on Iraq, hitting communications, power networks and airfields around Baghdad. Given the mission of finding and destroying Iraq's SCUD missile launchers as a result the squadron earned the nickname SCUD BUSTERS. The 335th FS made aerial warfare history by downing an Iraqi helicopter in the air using a laser guided bomb. During the war, the Chiefs flew 1,097 combat missions over Iraq and occupied Kuwait, dropping over 4.8 million pounds of ordnance. The 335th scored the F-15E's only air-to-air kill in Operation Desert Storm , when Capt. Richard Band and Daniel Bake (WSO) dropped a GBU-10 into a flying Iraqi Hughes 500 helicopter. The squadron still periodically deploys to Saudi Arabia to enforce the UN sanctions against Iraq.

F-15E 89-0483 is seen in this photograph, taken in April 1991, marked as the 335th FS Flagship. (Alec Fushi)

F-15E 88-1696 "aero-brakes" after landing at Nellis AFB in October 1993 following a Red Flag mission. (David F. Brown)

F-15E 90-0232 rotates for takeoff at Nellis AFB departing for a Red Flag mission in October 1993. (David F. Brown)

F-15E 88-1676 is seen here at Seymour-Johnson AFB on October 16, 1998. (Don Logan)

F-15E 89-0471 is seen here at Seymour-Johnson AFB on October 1997. (Alec Fushi)

F-15E 89-0502 is seen here at Seymour-Johnson AFB on October 16, 1998. (Don Logan)

336th FIGHTER SQUADRON
ROCKETEERS

The 336th Fighter Squadron was activated September 12, 1942, and like its sister squadrons, was made up mainly of American volunteers serving in the Royal Air Force prior to the United States' entry into World War II. Transitioning from the RAF duty as the 133rd Eagle Squadron, the 336th continued to fly British Spitfires until arrival of P-47 Thunderbolt fighter-bombers in 1943. About a year later the squadron transitioned to North American P-51 Mustangs, perhaps flying most renowned piston-engine fighter for the remainder of the year. By the war's end, the 336th destroyed 358 aircraft – 175 in the air and 183 on the ground with 21 pilots achieving ace status.

In April 1947, with the jet-powered F-80 Shooting Star, the 336th earned the name Rocketeers for its early association with the Air Corps transition to the jet age. In 1949, the 336th moved to Langley AFB, Va., with the F-86 Sabre jets. By November the Rocketeers were enroute to the Far East to fight the Korean Conflict. The 336th earned, along with its sister squadrons, the reputation of "MiG Killers" for actions along the Yalu River and other areas where combat was joined with Korean, Chinese and Russian pilots. On December 17, 1950, Lt. Col. Bruce Hinton, 336th commander, shot down a MiG-15 in the first ever all supersonic aerial combat. The Rocketeers were credited with 116.5 kills during the Korean Conflict, adding four aces to their rolls.

The 336th moved to Seymour-Johnson on December 3, 1947. The unit flew North American F-100 Super Sabres until mid-1959 when it transitioned to the Mach 2 capable Republic F-105 Thunderchief. In 1967 the Rocketeers began the McDonnell Douglas F-4D Phantom II, transitioning to the F-4E in July 1970. From April to September 1972 and again from March to September 1973, the 336th conducted Constant Guard operations from Ubon Royal Thai Air Base, Thailand, in support of Linebacker Operations in Southeast Asia. On August 15, 1972, while in Southeast Asia assigned to the 8th TFW, a 336th TFS F-4E was credited with destroying a North Vietnamese MiG 21 in aerial combat.

In October 1989 the 336th became the first operational F-15E Strike Eagle squadron in the Air Force when it achieved Initial Operational Capability (IOC) on 1st October 1, 1989. As the first F-15E squadron, the Rocketeers deployed in support of Operation Desert Shield on August 9, 1990. In December 1990, the 336th redeployed to Saudi Arabia in preparation for Operation Desert Storm. On January 16, 1991, the Rocketeers launched 24 aircraft against targets in Iraq to begin Operation Desert Storm and the liberation of Kuwait. The first night was an unqualified success as the fighting Rocketeers put their bombs on target and returned home safe and sound. By the end of Operation Desert Storm the 336th had flown 1,100 combat sorties, logging 3,200 hours and dropping 6.5 million pounds of ordnance on enemy targets, including a combination of general purpose, cluster and laser guided bombs. During the conflict, the squadron lost two aircraft, with one crew killed in the crash, and the other imprisoned by the Iraqis until the end of the war.

F-15E 90-0229 is seen in this photograph, taken on October 16, 1998, marked as the 336th FS Flagship. An AGM-130 missile is visible under the wing. (Don Logan)

F-15E 87-0182 is seen in this photograph, taken in April 1991, marked as the 336th TFS Flagship. (Alec Fushi)

F-15E 89-0483 is seen in this photograph, taken in December 1990, marked as the 336th AMU Flagship. (Alec Fushi)

F-15E 89-0505 is seen in this photograph, taken on May 19, 1995, marked as the 336th FS Flagship. (Norris Graser)

F-15E 89-0488, mismarked as 88-0488, is seen here at Seymour-Johnson AFB in May 1979. (Alec Fushi)

F-15E 88-1691 is seen here at Seymour-Johnson AFB in April 1991. Seven Desert Storm mission markings are visible under the windshield. (David F. Brown)

F-15E 90-0231 is seen here at Seymour-Johnson AFB on October 16, 1998. (Don Logan)

F-15E 87-0181 is seen here at Seymour-Johnson AFB in May 1997. (Alec Fushi)

F-15E 87-0195 is seen here at an Air Show stopover at McConnell AFB, Kansas in June 1997. (Don Logan)

F-15E 87-0207 is seen here at Seymour-Johnson AFB on October 16, 1998. Seven Desert Storm mission markings are visible under the windshield. (David F. Brown)

TYNDALL AFB, FLORIDA, 325th FIGHTER WING

1st, 2nd, 95th Fighter Squadrons Replacement Training Units for ACC and ANG F-15s.

The 325th Fighter Weapons Wing (FWW) began operations on July 1, 1981, as part of the Air Defense Weapons Center at Tyndall AFB, Florida. The 325th accomplished the operations, test and evaluation, and maintenance portions of the weapons center's mission, which was directly related to combat readiness training for air defense. The primary aircraft were the F-106 and T-33. On October 15, 1983 it was redesignated the 325th Tactical Training Wing, and assumed its air superiority training responsibilities as part of the Air Defense Weapons Center. The Wing also served as the Replacement Training Unit (RTU) for Air National Guard F-15 units, and was also responsible for all F-15 maintenance training. The Wing's first F-15A was received on December 7, 1983, and training of the first F-15A/B class began in August 1984. It was redesignated the 325th Fighter Wing on October 1, 1991. On July 1, 1993 the Wing was transferred to the Air Education and Training Command and the 19th Air Force. The Wing usually acted as the host for the bi-annual William Tell air-to-air weapons competition held at Tyndall. When the RTU at Luke AFB was disbanded, the 325th assumed RTU duties for all F-15 air-to-air training. In August 1991 the 325th was redesignated a Fighter Wing. Aircraft of the 325th Fighter Wing carry TY tail codes for TYndall AFB.

The 325th Fighter Wing's heritage goes back to the 325th Fighter Group Checkertail Clan of World War II fame. The 325th Fighter Group was activated August 3, 1942, at Mitchel Field, New York with P-40s. It entered combat April 17, 1943, escorting medium bombers, flying strafing missions and making sea sweeps from bases in Algeria and Tunisia. In May 1944, the unit transitioned to P-51 Mustangs which it flew until the end of World War II. By then the 325th's motto-Locare Et Liquidare (Locate and Liquidate) had earned the respect of the allies and Germans.

(Kevin Patrick)

Below: F-15C 84-0007 is seen here at Tyndall AFB, with the Air Education and Training Command (AETC) emblem on the tail, marked as the 325th Fighter Wing Flagship. (Alec Fushi)

After the 325th's return from Europe, it underwent a series of inactivations, reactivations and designations known at times as the 325th Fighter Group (All-Weather) and 325th Fighter Interceptor Group. Assigned aircraft included P-61s in 1947, F-82s in 1948 and F-94s in 1950. Then on August 18, 1955, the group was reactivated as the 325th Fighter Group (Air Defense), assigned to Air (later Aerospace) Defense Command, and equipped with F-86 Sabre jets. Its tactical units were the 317th and 318th Fighter Interceptor Squadrons.

On October 18, 1956, the Air Defense Command directed a Wing organization be set up at McChord Air Force Base, Washington and the 325th Fighter Wing (Air Defense) was activated. The unit's two tactical squadrons, the 317th and 318th Fighter Interceptor Squadrons, transitioned from the F-86 to the F-102A delta-wing, all-weather interceptor.

In August 1957, the 317th Fighter Interceptor Squadron was assigned to Alaskan Air Command, and the 325th gained the 64th Fighter Interceptor Squadron from Alaska. Early in 1960, the 325th

Fighter Group began a gradual phase-out of the F-102 and a transition to the F-106 Delta Dart. Before the completion of this transition, the 325th was inactivated on March 25, 1960.

The 325th was activated at Tyndall in July 1981 as the 325th Fighter Weapons Wing. The wing accomplished the operations, test and evaluation and maintenance portions of the complex United States Air Force Defense Weapons Center mission, which was directly related to combat readiness training for air defense. The primary aircraft were the F-106 and T-33.

On October 15, 1983, it was redesignated the 325th Tactical Training Wing, and assumed its air superiority training responsibilities as part of the United States Air Force Air Defense Weapons Center. The wing began receiving the F-15 Eagle in April 1984. When the United States Air Force Air Defense Weapons Center was inactivated September 12, 1991, the 325th Tactical Training Wing assumed the role as installation host. It was redesignated the 325th Fighter Wing October 1, 1991.

The 1st TFTS Flagship and 2nd TFTS Flagship, accompanied by unit T-33 are being led by the 325th TTW Flagship in this photo taken in Summer 1986. (Boeing)

F-15A 75-0057 is seen here at Nellis AFB in August 1990, marked as the 325th Tactical Training Wing Flagship. (Ben Knowles)

F-15C 83-0022 is seen on May 21, 1993 marked as the 325th Fighter Wing Flagship. (Keith Snyder)

F-15A 75-0082 is seen here at Tyndall AFB, with the Air Combat Command (ACC) emblem on the tail, marked as the 1st Air Force Flagship. (David F. Brown)

F-15D 82-0044 is seen here at Tyndall AFB, with the Air Education and Training Command (AETC) emblem on the tail, marked as the 19th Air Force Flagship. (Alec Fushi)

1st FIGHTER SQUADRON
FIGHTIN FURIES and GRIFFINS

Known as the Fightin Furies, the history of the First and Finest is long and distinguished. Beginning in World War II, the 1st Fighter Squadron saw service in the Pacific Theater, where it flew the P-47 Thunderbolt. During this era the squadron emblem was Miss Fury, a mythical Greek goddess of vengeance, wearing a black form-fitting gown, a cape, and boots, and sitting on a white cloud, holding a skull. The 1st FS was inactivated in October 1946 at the conclusion of World War II. From 1954 to 1959 the 1st operated out of George AFB, California flying F-86s and F-100s until it was again inactivated on March 15, 1959.

In January 1984, the squadron was reactivated as the 1st Tactical Fighter Training Squadron (Griffins) as part of the 325th Tactical Training Wing, at Tyndall AFB, Florida to train Air Force pilots in the F-15 Eagle. As the Griffins, a new patch was designed. On September 17, 1991, the squadron was renamed the 1st Fighter Squadron, and once again adopted the "Miss Fury" emblem.

F-15C 84-0017 is seen here at Tyndall AFB, with the Air Education and Training Command (AETC) emblem on the tail, marked as the 1st Fighter Squadron Flagship. (Alec Fushi)

F-15A 74-0101, marked as the 1st TFTS, is seen over the Gulf of Mexico during the Summer 1986. (Boeing)

F-15AC 74-0101 is seen here at Luke AFB on December 4, 1985, with the Air Combat Command (ACC) emblem on the tail, marked as the 1st Tactical Fighter Training Squadron (TFTS) Flagship. (Kevin Patrick)

F-15C 74-0101 is seen here at Luke AFB in March 1986 marked as the 1st Fighter Squadron Flagship. (Douglas Slowiak/Vortex Photo Graphics)

F-15C 78-0505 is seen taxiing at Tyndall AFB on October 3, 1993, with the Air Education and Training Command (AETC) emblem on the tail, marked as the 1st Fighter Squadron Flagship. (Alec Fushi)

F-15A 74-0103 with the red stripe of the 1st TFTS and the ACC emblem is seen here at Luke AFB in March 1986. (Douglas Slowiak/Vortex Photo Graphics)

F-15D 78-0573 with the red stripe of the 1st FS and the AETC emblem is seen here at Tyndall AFB in January 1996. (Alec Fushi)

F-15A 74-0114 with the red stripe of the 1st TFTS and the ACC emblem is seen here at Luke AFB in March 1986. (Douglas Slowiak/Vortex Photo Graphics)

F-15C 83-0030 with the red stripe of the 1st FS and the AETC emblem is seen here at Tyndall AFB in December 1996. (Alec Fushi)

2nd FIGHTER SQUADRON
SECOND TO NONE, UNICORNS, or HORNEY HORSES

The 2nd Fighter Squadron was activated on January 15, 1941, as the 2nd Pursuit Squadron, it served in World War II with the 52nd Pursuit Group and flew the Curtiss P-40 Warhawk and Bell P-39 Airacobra. The Squadron also flew combat operations in the Supermarine Spitfire and the P-51 Mustang in the European and Mediterranean theaters and proudly boasts 11 fighter aces from that era. It was inactivated on November 7, 1945. Activated again on 9 Nov 1946, it was assigned to the 52nd Fighter Group and served tours in Schweinfurt and Bad Kissingen, Germany. It returned to Mitchel Field, New York and was redesignated the 2nd Fighter Squadron (All Weather) flying the Northrop P-61 Black Widow. In 1949, the squadron was moved to McGuire Field, New Jersey. Its assigned aircraft changed to the North American F-82 Twin Mustang.

In 1950, the 2nd became the 2nd Fighter All Weather Squadron. One year later the unit was redesignated the 2nd Fighter Interceptor Squadron and assigned the Republic F-84 Thunderjet. Realignment in 1952 saw the 2nd assigned first to the 4709th Defense Wing, followed one year later by transfer to the 568th Air Defense Group. In 1953, the 2nd was assigned the North American F-86D Sabre. In August 1995, the unit was reassigned to the 52nd Fighter Group (later Wing) and moved its operations to Suffolk County AFB, New York. 1957 saw the first of the delta-wing fighters assigned to the 2nd with the introduction of the Convair F-102 Delta Dagger. In 1959, the F-102 was replaced by the McDonnell F-101 Voodoo which flew with the squadron until 1969, when the squadron was inactivated. In 1971, the 2nd was again reactivated in the 23rd Air Division at Wurtsmith AFB, Michigan flying the F-106 Delta Dart. It was again inactivated March 31, 1973.

It was activated in August, 1974 as the 2nd Fighter Interceptor Training Squadron at the Air Defense Weapons Center at Tyndall AFB, again flying the F-106 Delta Dart. On February 1, 1982, the 2nd was redesignated the 2nd Fighter Weapons Squadron. On May 1, 1984 the unit redesignated as the 2nd Tactical Fighter Training Squadron. The squadron's mission was to train F-15 pilots. On November 1, 1991, the unit was redesignated the 2nd Fighter Squadron.

Below: F-15C 81-0045 is seen here at Tyndall AFB, with the Air Education and Training Command (AETC) emblem on the tail, marked as the 2nd Fighter Squadron Flagship. (Alec Fushi)

F-15B 77-0157 with the yellow stripe of the 2nd TFTS and the ACC emblem is seen here at Tinker AFB in March 1986. (Jim Geer)

F-15C 78-0535 with the yellow stripe of the 2nd FS and the AETC emblem is seen here at Luke AFB in October 1993. (Ben Knowles)

F-15D 78-0570 with the yellow stripe of the 2nd FS is seen here at Tyndall AFB in January 1996. (Alec Fushi)

F-15C 82-0032 with the yellow stripe of the 2nd TFTS and the ACC emblem is seen here at Tyndall AFB in December 1996. (Alec Fushi)

F-15D 82-0048 with the yellow stripe of the 2nd FS and the AETC emblem is seen here at Tyndall AFB in October 1, 1993. (Gilles Alleuard)

F-15D 80-0060 with the yellow stripe of the 2nd FS is seen here at Tinker AFB in August 1994. (Jim Geer)

95th FIGHTER SQUADRON
BONEHEADS

Known proudly as the Boneheads, the squadron's history began in 1942. The squadron first saw service flying the twin-tailed P-38 Lightning. The 95th FS finished the war with more than 400 kills, including 199 aerial victories. "Mr. Bones" is pictured on the unit patch with top hat, monocle, and cane. The 95th FS had previously been designated a Fighter Interceptor Training Squadron (FITS), but was redesignated shortly after it began receiving F-15A/Bs in April 1988.

F-15C 81-0048 is seen here at Tyndall AFB, in December 1996 with the Air Education and Training Command (AETC) emblem on the tail, marked as the 95th Fighter Squadron Flagship. (Alec Fushi)

Above: F-15A 75-0037 is seen here at Nellis AFB in August 1990, with the Air Combat Command (ACC) emblem on the tail, marked as the 95th Tactical Fighter Training Squadron Flagship. (Ben Knowles)

Right: F-15D 78-0572 with the blue stripe of the 95th FS and the ACC emblem is seen here at Tyndall AFB in January 1996. (Alec Fushi)

Below: F-15C 84-0022 with the blue stripe of the 95th FS and the AETC emblem is seen here at Tyndall AFB in December 1996. (Alec Fushi)

F-15B 73-0113 with the blue stripe of the 95th TFTS and the ACC emblem is seen in June 1991. (David F. Brown)

F-15A 75-0038 in 95th TFTS markings is seen here at Luke AFB in June 1990. (David F. Brown)

F-15B 77-0162 with the blue stripe of the 95th TFTS is seen here at Luke AFB in June 1990. (David F. Brown)

PACIFIC AIR FORCES/ALASKAN AIR COMMAND UNITS

ELMENDORF AFB, ALASKA

3rd Wing - 19th (F-15C) 43rd (F-15A and F-15C), 54th (F-15C), 90th (F-15E) Fighter Squadrons
Originally designated the 21st Composite Wing

Below: F-15A 74-0092 of the 21st TFW goes vertical in September 1983. (Boeing)

21st TACTICAL FIGHTER WING

Originally established at Elmendorf AFB, Alaska, in 1966 and assigned to the Alaskan Air Command as the 21st Composite Wing, the unit was assigned the air defense of Alaska and adjoining parts of Canada using F-4Es assigned to the 43rd TFS. In addition, the 5021st Tactical Operations Squadron (TOS) flew T-33 for target support operations. In 1977, the Wing added another F-4E squadron, the 18th TFS, with a primary air-to-ground mission. The 18th was assigned under the Wing's 343rd Tactical Fighter Group from November 15, 1977, until January 1, 1980. As part of an effort to modernize Alaskan Air Command air forces, the Wing began converting from F-4Es to F-15s in 1982, with the first aircraft arriving at Elmendorf AFB on March 1, 1982. The 43rd assumed air defense alert with the F-15A on October 5, 1982. A second F-15 squadron, the 54th, was added in 1987, and shortly thereafter both squadrons converted to newer F-15C/D models equipped with conformal fuel tanks for extended range operations. The Alaskan Air Command became the Alaska Command of PACAF on July 7, 1989. The Alaskan Air Command was redesignated Eleventh Air Force (11 AF), on August 9, 1990. This made the 21st TFW a PACAF Wing. The 21st TFW was redesignated as a 21st Wing on September 26, 1991. This designation was short-lived. In a move to preserve the Air Force's most illustrious units, the unit was inactivated being replaced by the 3rd Wing on December 19, 1991. For the same reason, the 21st returned to active duty on May 15, 1992, at Peterson AFB, Colorado as the non-flying 21st Space Wing assigned to Space Command.

Below: F-15A 74-0100 is seen here at Luke AFB on February 15, 1987 marked as the 21st TFW Flagship carrying only the AK tail codes and the special unit number above the serial number. (David F. Brown)

(Douglas Slowiak/Vortex Photo Graphics)

Right: F-15A 74-0100 is seen on November 16, 1985 marked as the 21st TFW Flagship. In addition to the special tail markings, 0100 also carried a map of Alaska overlaid with "Anchorage" under the cockpit. (Douglas Slowiak/Vortex Photo Graphics)

F-15C 81-0021 is seen here at Nellis AFB on January 30, 1988 marked as the 21st TFW Flagship. (Kevin Patrick)

F-15E 88-0667 is seen here at Luke AFB on May 16, 1991 marked as the 21st TFW Flagship. This was an unofficial marking, as the 21st TFW never had F-15E aircraft assigned. By the time the 90th FS had received its F-15Es the 21st TFW had changed to the 3rd Wing. (Kevin Patrick)

3rd WING

The 3rd Wing had previously been based at Clark AB in the Philippines and was moved when a lease agreement could not be reached with the island government. The Wing uses AK tail codes, signifying AlasKa. Originally the Wing's air superiority F-15 aircraft were painted in Compass Ghost with a Big Dipper and 'North Star' painted on a dark field inside each vertical stabilizer. The dark field has disappeared from the aircraft as they have been repainted into the Mod Eagle scheme. Currently, the inside of the vertical tails have a black Big Dipper and North Star painted over the standard camouflage.

Trained as a bombardment and reconnaissance Wing prior to Korean War. The Wing performed reconnaissance and interdiction combat missions from Iwakuni AB, Japan at the beginning of the Korean War, July 1 through 19, 1950. From July 20 to December 1, 1950, the tactical group and its squadrons served under operational control of another organization. The Wing assumed a supporting role, initially from Johnson AB, Japan, but later from Yokota, Japan. The Wing returned to Iwakuni AB on December 1, 1950, regained control of its combat units and performed night intruder combat missions. It moved to South Korea in August 1951 and interdicted main supply routes in western North Korea for the remainder of the war. After the Korean war, the Wing participated successively in bombardment, air defense, reconnaissance, and air refueling training. Its headquarters was non-operational September 1, 1963 to January 8, 1964. It moved to the United States without personnel or equipment in January 1964, then trained and rotated its squadrons in detached status to Southeast Asia for combat duty. It moved in November 1965 to Bien Hoa AB. South Vietnam, a forward operating base which frequently came under enemy mortar and rocket fire. Missions included close air support, counterinsurgency, forward air control, interdiction, and radar-controlled bombing. It supported numerous ground operations with strike missions against enemy fortifications, supply areas, lines of communication and personnel, in addition to suppressing fire in landing areas. During this time, the Wing also participated in combat evaluation of the F-5 and A-37 aircraft. Unmanned and unequipped on October 31, 1970, the Wing remained active in a "paper" status until it moved to South Korea on March 15, 1971, where it was manned and equipped with F-4 aircraft. In September 1974, it moved without personnel or equipment to Clark AB, Philippines, replacing the 405th Fighter Wing. It participated in frequent operational exercises and evaluations. Between April 5 and May 31, 1975, the Wing used its facilities as a staging area for Operations Baby Lift (evacuation of Vietnamese orphans from South Vietnam to the United States) and New Life (evacuation of Vietnamese adults to the United States for resettlement). The Wing performed fighter aggressor training operations using T-38 and later F-5E aircraft from 1976 to 1988 and deployed throughout Pacific Air Forces to provide dissimilar aircraft combat training to US and allied fighter units.

(Douglas Slowiak/Vortex Photo Graphics)

It deployed aircraft from the Philippines to Korea annually to participate in multinational joint-service combined forces exercises, 1978-1991. With addition of the F-4G Wild Weasel aircraft in 1979, the Wing acquired dual role capabilities of air-to-air, air-to-ground, and defense suppression/electronic countermeasures. Assignment of the 1st Special Operations Squadron, equipped with MC-130 aircraft, provided the Wing with an unconventional warfare capability, January 1981 through March 1983. It operated UH-IN helicopters, 1988-1991, for drone recovery, VIP airlift, range support, Philippine air defense site support, and medical evacuation. At the end of May 1991, the last F-4 aircraft departed the Wing, shortly before the eruption of Mount Pinatubo in June 1991. The Wing was not operational from June 1991 until it moved on paper to Elmendorf AFB on December 19, 1991, replacing the 21st Tactical Fighter Wing. The Wing expanded the air defense mission of Alaska with the F-15E aircraft to include deep interdiction and air-to-air capabilities.

F-15C 79-0020 is seen here at Luke AFB on October 18, 1997 marked as the 3rd Wing Flagship. (Douglas Slowiak/Vortex Photo Graphics)

F-15E 90-0243 is seen on the runway at Nellis AFB in October 1993 marked as the 3rd Wing Flagship. (David F. Brown)

11th AIR FORCE

Right and below: F-15C 81-0021 is seen here at Luke AFB on December 2, 1995 marked as the 11th Air Force Flagship. In addition to the special tail markings, 0021 also carried a map of Alaska overlaid with "Anchorage" under the cockpit. (Kevin Patrick)

F-15C 85-0103 is seen here at Hickam AFB, Hawaii on April 5, 1999 marked as the 3rd Operations Group (OG) Flagship. (Robert F. Dorr)

ALASKA 49th STATE FLAGSHIP

F-15C is seen in these two photos marked as the Alaska (The 49th State of the U.S.). Flagship in both camouflage schemes. (Don Logan ??)

19th FIGHTER SQUADRON
GAMECOCKS

The 19th Fighter Squadron was known as the 19th Aero Service Squadron and was stationed in Texas, Ohio, New York and France. Originally established as an Army Flying School Squadron, the 19th was based at what is now known as Kelly Air Force Base, Texas. The squadron subsequently moved to Ohio and New York for short periods before ending up at Clermont, France to observe the French company Michelin's airplane manufacture/assembly procedures.

Following World War I, the 19th Pursuit Squadron spent the next few years of its existence at various locations in the Hawaiian Islands, flying and training in M-A and SE-SA aircraft at first, then moving on to P-26s, P-40s and P-47s. The 19th suffered six casualties as a result of the infamous attack on Oahu by the Japanese on December 7, 1941, but luckily had no fatalities. Oddly enough, the 19th had been scheduled to relocate to the Philippines on December 1, but their orders were changed by higher headquarters to an December 8 departure.

The 19th Fighter Squadron, flying P-47D Thunderbolts, was given its first wartime tasking and moved to Natoma Bay, Saipan on April 18, 1944. The unit was inactivated on January 12, 1946. Redesignated the 19th Tactical Fighter Squadron the squadron was activated at Shaw AFB, South Carolina and assigned General Dynamics F-16s. It was inactivated on December 31, 1993. It was reactivated on January 1, 1994, replacing the 43rd Fighter Squadron, taking over their F-15C and D air superiority fighters. The 19th FS flew 160 sorties during Operation Provide Comfort.

Below: F-15C 79-0079 is seen landing at Elmendorf AFB marked as the 19th Fighter Squadron Flagship. (USAF)

227

F-15C 78-0546 is seen here at Luke AFB on October 18, 1997. (Douglas Slowiak/Vortex Photo Graphics)

F-15D 78-0564 is also seen at Luke AFB on October 18, 1997. (Douglas Slowiak/Vortex Photo Graphics)

F-15C 80-0020 is seen here at Luke AFB on October 18, 1997. (Douglas Slowiak/Vortex Photo Graphics)

43rd FIGHTER SQUADRON
BUMBLE BEES

The 43rd TFS was assigned air defense of Alaska using F-4Es on July 15, 1970. The Wing began transitioning from F-4Es to F-15s in 1982, with the 43rd TFS receiving their first F-15A/B on June 21, 1982.

The 43rd assumed air defense alert with the F-15A on October 5, 1982. The unit transitioned into F-15C/Ds previously used by the 1st TFW at Langley AFB, Virginia beginning on May 23, 1987. The 43rd TFS was redesignated the 43rd FS on September 26, 1991. It was inactivated on December 31, 1993, and was operationally replaced by the 19th Fighter Squadron, an F-16 squadron which moved from Shaw AFB, South Carolina on January 1, 1994. The 19th FS took over the 43rd's F-15C/Ds.

F-15C 80-0043 is seen here at Luke AFB in July 1990 marked as the 43rd TFS/AMU Flagship. (Ben Knowles)

F-15C 81-0043 is seen here at in September 1992 marked as the 43rd FS Flagship. (Ben Knowles)

F-15A 74-0115 is seen in October 1986 while assigned to the 43rd TFS, 21st TFW. (Jim Goodall)

F-15C 78-0549 is seen here at Luke AFB on July 22, 1990. (Kevin Patrick)

F-15C 79-0075 is seen on the ramp at Elmendorf AFB in August 1993. (Alec Fushi)

F-15D 79-0009 is seen here at the London, Ontario Air Show in June 1991. (David F. Brown)

F-15C 79-0054 is seen here at Elmendorf AFB in July 1989. The aircraft carries a non-standard serial number presentation. (Jim Goodall)

F-15B 74-0138 is seen here at Luke AFB on October 5, 1996. (Kevin Patrick)

F-15C 79-0059 is seen on the ramp at Elmendorf AFB in August 1993. (Alec Fushi)

This 3rd Wing four-ship, aligned by serial number, is seen flying over the mountains of the Alaska Range in September 1983. (Boeing)

54th FIGHTER SQUADRON
FIGHTING 54TH or LEOPARDS

The Fighting 54th was activated as the 54th Pursuit Squadron (Interceptor) on January 15, 1941 at Hamilton Field as part of the 55th Pursuit Group. The 54th began operational training using a motley assortment of aircraft, the P-36 Hawk, P-40B Tomahawk, and rare P-43 Lancers. The squadron received its first P-38 Lightning on January 13, 1942. On May 23, 1942, the squadron's aircraft began a long ferry mission to Alaska where they were to serve with distinction throughout World War II. Anticipating Japanese moves against the Aleutian Islands, the squadron arrived at Elmendorf Field, Fort Richardson, Alaska, on June 2, 1942.

The 54th demobilized at Ft. Lawton, Washington on March 21, 1946. During the cold war, the squadron activated at Rapid City Air Force Base, South Dakota (later Ellsworth AFB), as the 54th Fighter Interceptor Squadron (FIS), part of the Air Defense Command in 1952. Flying F-51s, F-84s, and F-89s, the 54th remained active until 1960, providing air defense for the north and central United States. In 1959, flying F-89J Scorpions, the squadron won the Hughes Trophy for outstanding air defense. Despite this success, the Air Force targeted the 54th for inactivation the following year as the Scorpion was phased out of front line service. After being inactive for 27 years, having last flown F-89Js in 1960, a resurgence of Cold War tensions and the Soviet long range bomber/cruise missile threat caused the 54th to be activated again on May 8, 1987, at Elmendorf AFB as an additional air defense squadron. Initially flying F-15A/Bs

Eagles, the squadron converted to the newer F-15C/Ds the following year. The 54th TFS was redesignated the 54th FS September 26, 1991, and was transferred to 3rd Wing on December 19, 1991.

The squadron participated in Operation Provide Comfort from April through June 1995, with significant support provided by personnel from the 19th FS. On April 21, 1999, in support of Operation Allied Force the 54th FS deployed six F-15C aircraft to RAF Lakenheath. The pilots and ground personnel continued on to Cervia Air Base, Italy and were assigned to the 493rd Expeditionary Fighter Squadron which had deployed to Cervia AB from RAF Lakenheath. The aircraft remained at RAF Lakenheath and were used for training of 493rd pilots who had not deployed.

F-15C 81-0054 is seen here at Luke AFB on October 18, 1997 marked as the 54th Fighter Squadron Flagship. (Kevin Patrick)

F-15C 81-0054 is seen here at Nellis AFB in November 1989 marked as the 54th TFS/AMU Flagship. (Alec Fushi)

Above: F-15C 80-0053 is seen here at Nellis AFB on April 18, 1990 marked as the 54th TFS/AMU Flagship. (Brian C. Rogers)

Right: F-15D 78-0574 is seen here at Luke AFB on March 29, 1992. (Kevin Patrick)

F-15C 81-0020 is seen here at Nellis AFB in November 1989. (Alec Fushi)

F-15D 80-0058 is seen here at Tinker AFB in September 1992. (Jim Geer)

F-15C 81-0021 is also seen at Tinker AFB in September 1992. (Jim Geer)

F-15C 80-0002 is seen here at Luke AFB on December 2, 1995. (Kevin Patrick)

90th FIGHTER SQUADRON
PAIR O'DICE

The 90th Pair O'Dice Fighter Squadron was activated in early 1917 as the 90th Aero Squadron. The 90th Aero Squadron originated at Kelly Field, Texas. The 90th arrived in Liverpool, England on November 10, 1917, and shortly left for Colombey les Belies, France. The squadron returned to the United States on April 20, 1919, and eventually moved back to Kelly Field, Texas. The squadron continued to train as an attack squadron. In 1936 the squadron became part of the 3rd Bombardment Group flying North American B-25 Mitchell bombers. The squadron embarked for Australia in February 1942 and led the way in the Pacific Theater. In 1945 the squadron moved to Okinawa in preparation for a major offensive on Japan, but the war ended with the Japanese surrender. The squadron then assumed occupational duty and moved to Yokota Air Base, Japan, before being inactivated on October 1, 1949. The 90th Fighter Squadron was reactivated on June 25, 1951, at Iwakuni Air Base, Japan. During August 1951, the squadron moved to Kunsan Air Base, South Korea after the Allied forces gained a stronger foothold on the peninsula.

As the United States became involved with Vietnam, the 90th changed its name, yet again, to the 90th Special Operations Squadron. The squadron flew F-100's, A-37B's, and MC-130's and was stationed at three different bases from 1966 to 1972. After the Vietnam conflict ended, the squadron returned to Clark AB, Philippines. As Desert Storm arose, the 90th Fighter Squadron sent six F-4G's to Incirlik Air Base, Turkey. The squadron's involvement in Desert Storm gave the 90th the distinction of being the only squadron in the Pacific Air Forces to have participated in all five wars in which the US air forces were involved.

On May 29, 1991 the 3rd Wing acquired the 90th TFS from Clark AFB, moved the 90th to Elmendorf AFB, Alaska, and replaced its F-4Es with F-15E Strike Eagles. The 90th TFS was redesignated the 90th FS September 26, 1991, and was transferred to the 3rd Wing on December 19, 1991.

F-15E 90-0233 is seen here at Luke AFB on March 29, 1992 marked as the 90th Fighter Squadron Flagship. (Kevin Patrick)

A four-ship formation of 90th FS F-15Es is seen flying near Mt McKinley, Alaska. (Boeing)

Right: F-15E 87-0210, mismarked as 88-0210, is seen here at Luke AFB on May 16, 1991 marked as the 90th TFS/AMU Wing Flagship. (Kevin Patrick)

Below: F-15E 90-0233 is seen in August 1997 marked as the 90th FS Flagship. (Alec Fushi)

F-15E 90-0233 is seen here at Luke AFB on October 18, 1997 marked as the 90th FS Flagship. (David F. Brown)

F-15E 90-0233 is seen landing at Nellis AFB in October 1993 marked as the 90th FS Flagship. (David F. Brown)

F-15E 90-0238 is seen on February 7, 1993. (Don Logan Collection)

This 90th FS F-15E is seen landing at Nellis AFB in October 1993. (David F. Brown)

F-15E 90-0245 is seen on a stopover at Luke AFB in October 1993. (Ben Knowles)

F-15E 90-0246 is seen here at Luke AFB on January 16, 1993. (Kevin Patrick)

F-15E 78-0250 is seen in June 1992. (Tony Cassanova)

KADENA AIR BASE, OKINAWA, JAPAN, 18th WING

12th, 44th, 67th Fighter Squadrons

The 18th Wing is based at Kadena AB in Okinawa, Japan, and consists of the 12th FS, 44th FS, and 67th FS. The Wing is under the control of the Pacific Air Forces (PACAF), Fifth Air Force. The Wing received its first F-15C/D on September 26, 1979, replacing F-4Ds. The F-15s had previously been operated by the 33rd FW at Eglin AFB prior to reaching Kadena. Due to the remote operating location, most depot level maintenance is performed under contract by Korean Air Lines in Pusan, Korea. On a rotating basis, the 18th FW provides aircraft to stand alert on the Korean peninsula. During the 1980s, in a slight variation of the standard Air Force tail stripes, every aircraft in the 18th FW carried all three squadron colors. The outermost color denotes the squadron to which the aircraft is assigned. This multi-color tail stripe was later replaced by a stripe of only the squadron color. Most aircraft have carried a stylized shogun warrior on the inside of each tail, and the ZZ tail codes do not represent anything in particular. The 18th Wing has F-15C/D, KC-135R, E-3 AWACS and HH-60 aircraft assigned.

The 18th Wing is the Kadena Air Base host unit. It is the oldest Wing on continuous active duty and is the only active Wing whose fighters have never served in the continental United States. From December 1, 1948 to May 16, 1949, was the major Far East Air Forces organization in the Philippines. In late July 1950, the group and two squadrons deployed to Korea for combat, converting to F-51s, while the Wing continued air defense of the Philippines. The Wing rejoined the group in Korea on December 1, 1950 and resumed operational control. Combat operations included armed reconnaissance, strategic bombing, close ground support, aerial combat and interdiction. It converted to F-86s in early 1953 and continued counter-air and ground attack missions to the end of the war. After the close of the Korean conflict, the Wing transferred to Okinawa on November 1, 1954, supporting tactical operations there and in Korea, Japan, Formosa (later Taiwan), and the Philippines. the 18th supported combat operations in Southeast Asia from 1961 with deployed reconnaissance forces and from 1964

(Boeing)

with deployed tactical fighter forces until the end of that conflict. It deployed at Osan AB, South Korea, following the Pueblo crisis (January 28 through June 13, 1968). It maintained an air defense alert capability in South Korea beginning in 1978. The 18th converted from F-4 to F-15 aircraft during 1979 and 1980, and continued to maintain assigned aircraft, crews, and supporting personnel in a high state of readiness for tactical air requirements of Fifth Air Force and the Pacific Air Forces. Beginning October 1991, the mission of the Wing expanded to include aerial refueling (KC-135R) and surveillance, warning, command, control and communications (E-3 AWACS). In February 1993, the 18th gained responsibility for coordinating rescue operations in the Western Pacific and Indian Ocean.

The 18th Operations Group conducts the Wing's flying mission. The group incorporates three fighter squadrons flying F-15C/D Eagles, the 909th Air Refueling Squadron with KC-135Rs, the 961st Airborne Air Control Squadron flying E-3B/C aircraft, the 33rd Rescue Squadron with HH-60 helicopters, the 623d Air Control Flight, and the 18th Operations Support Squadron. The three assigned F-15 Fighter Squadrons; the 12th, 44th, and 67th have the primary mission of air superiority.

F-15C 78-0518 is seen here at William Tell 96 at Tyndall AFB in October 1996 marked as the 18th Wing Flagship. (Alec Fushi)

F-15C 78-0518 is seen in July 1988 marked as the 18th Tactical Fighter Wing Flagship. (Don Logan Collection)

F-15C 78-0479 is seen here at Kadena AB in December 1995 marked as the 18th Wing Flagship. (Alec Fushi)

F-15C 78-0479 is seen here at William Tell 96 at Tyndall AFB on October 23, 1996 marked as the 18th Wing Flagship. (Nate Leong)

F-15C 78-0474 is seen here at William Tell 88 at Tyndall AFB on October 11, 1988. (Don Logan Collection)

Above: F-15C 78-0480 is seen here at William Tell 86 at Tyndall AFB on October 23, 1986. (Peter Wilson)

Right: F-15C 78-0531 is seen here at William Tell 94 at Tyndall AFB in October 1994. (David F. Brown)

F-15C 78-0541 is seen here at Kadena AB in July 1988. (Don Logan Collection)

F-15C 78-0498 is seen here at William Tell 96 at Tyndall AFB in October 1996. (Alec Fushi)

12th FIGHTER SQUADRON
DIRTY DOZEN

The 12th Pursuit Squadron was activated at Selfridge Air Force Base, Michigan, in January 1941. The unit and its P-39 Airacobras were shipped to the Pacific Theater in early 1942. In 1943, the unit converted to P-38 Lightnings and in April, became part of the 18th Fighter Group. The 12th Tactical Fighter Squadron (TFS) was assigned to Guadalcanal during the spring and summer of 1943. During this period, pilots of the 12th TFS participated in a special mission to the Shortland Islands in which a bomber carrying Admiral Yamamoto was shot down. The last daylight raid against Guadalcanal came in June 1943 when 68 enemy bombers and 50 fighters hammered friendly positions. The 12th was active during this battle in which 20 enemy aircraft were shot down. The "Dirty Dozen", as the unit came to be known, continued operations against the Shortland Islands through late 1943. During the last eight months of World War II, the 12th TFS was based on Luzon. Following the cessation of hostilities, the unit remained in the Philippines, continuing to upgrade its combat potential with the addition of the P-51 Mustang in 1946. The 12th TFS took part in combat in Korea from July 28, 1950 through July 27, 1953 flying the F-80 and F-86.

The 12th moved to Kadena AB on October 30, 1954, still flying F-86s. In 1957 the 12th traded its F-86s for F-100s, and in 1963 traded its F-100s for F-105s. The F-105s were replaced with F-4Ds. The 12th started receiving F-15C/Ds in 1980. The 12th TFS was redesignated the 12th FS on October 1, 1991. The aircraft of the 12th FS wear ZZ tail codes and a yellow tail stripe.

(Jerry Geer)

F-15C 78-0512 is seen here at Kadena AB in December 1995 marked as the 12th Fighter Squadron (FS) Flagship and carrying the yellow tail stripe of the 12th FS. (Alec Fushi)

F-15C 78-0477 is seen here at Kadena AB in December 1995 carrying the yellow tail stripe of the 12th FS. (Alec Fushi)

F-15C 78-0491 is seen here at Kadena AB in December 1995 in 12th FS markings. (Alec Fushi)

F-15C 78-0511 is also seen at Kadena AB in December 1995 carrying the yellow tail stripe of the 12th FS. (Alec Fushi)

44th FIGHTER SQUADRON
VAMPIRES

The 44th Pursuit Squadron was based at Bellows Field on Oahu, Territory of Hawaii on December 7, 1941 and took part in the defense of the Hawaiian Islands following the Japanese attack. They remained on Oahu flying combat patrols until October 1942. On December 20, 1942 they arrived at Guadalcanal and remained in the Southwestern Pacific for the rest of World War II. The 44th moved to Kadena AB on July 15, 1955, flying F-86s. In 1957 the 44th traded its F-86s for F-100s, and in 1963 traded its F-100s for F-105s. The F-105s were replaced with F-4C and in 1975 with F-4Ds. The 44th started receiving F-15C/Ds in 1979. In October 1991 the 18th Tactical Fighter Wing went through a reorganization and a name change and the 44th Tactical Fighter Squadron was redesignated the 44th Fighter Squadron. In 1995, the Vampires selected their first ever unit motto: "Desmodus Vinco Umbique", which means Vampire Bats Prevail Anywhere

From mid-1996 to early 1997, the 44th upgraded their F-15 with Night Vision Imaging System (NIVS), becoming the first active duty Air Force Squadron with NIVS. This modification allows 44th fighter pilots to use the Night Vision Goggles without being blinded by the bright white lights of the cockpit at night. The aircraft of the 44th FS wear ZZ tail codes and a black tail stripe.

F-15C 78-0544 is seen here at Kadena AB in December 1995 marked as the 44th Fighter Squadron (FS) Flagship and carrying the black tail stripe of the 44th FS. (Alec Fushi)

F-15C 78-0493 is seen here at a stopover at Luke AFB on May 24, 1992 carrying the black tail stripe of the 44th FS. (Kevin Patrick)

F-15C 78-0496 is seen in December 1995 carrying the black tail stripe of the 44th FS. (Alec Fushi)

F-15C 78-0509 is seen here at Kadena AB in December 1995 carrying 44th FS markings. (Alec Fushi)

67th FIGHTER SQUADRON
FIGHTIN' COCKS

The 67th Pursuit Squadron was activated at Selfridge Field, Michigan, on January 15, 1941. The squadron was deployed to Camp Darly, Australia, on January 23, 1942. On May 15, 1942 the 67th Pursuit Squadron was redesignated the 67th Fighter Squadron. During World War II, it flew P-39 Airacobra and P-38 Lightning aircraft. From the end of World War II until the outbreak of the hostilities in the Republic of Korea, the 67th Fighter Squadron remained in the Republic of the Philippines. The squadron flew its first combat mission in the Republic of Korea on August 1, 1950. From the end of the Korean Conflict until 1953, the 67th Fighter-Bomber Squadron flew the P-51 Mustang and the P-80 Shooting Star. In December 1953, the squadron was equipped with the F-86 Sabre. In 1954, the 67th moved to Kadena Air Base, Japan and transitioned to the F-100 Super Sabre in 1957. On 1 July 1, 1958, the squadron was renamed the 67th Tactical Fighter Squadron. In 1963, the 67th Tactical Fighter Squadron became the first squadron in Pacific Air Forces to receive the F-105D Thunderchief. Beginning in January 1965, the 67th Tactical Fighter Squadron deployed crews on temporary duty to Southeast Asia to help with the build-up of aircraft and aircrews. Between February 1965 and November 1967, the squadron deployed to Korat Royal Thai Air Force Base, Thailand, where it flew combat missions in the F-105.

From 1967 to 1971, the 67th Tactical fighter Squadron transferred to Misawa Air Base Japan, where it converted to the F-4C Phantom II, and was tasked with air defense of Japan and South

Korea. In 1971, the 67th returned to Kadena AB, but staged out of Korat, Thailand, using the F-4C Wild Weasel in combat.

The 67th Tactical Fighter Squadron converted to the F-15 Eagle on September 29, 1979, changing their mission to air superiority. The 67th Tactical Fighter Squadron was the first squadron in Pacific Air Forces to fly the F-15. On October 1, 1991 Tactical was dropped from the designation and returned to the earlier designation as the 67th Fighter Squadron. From October 1996 through January 1997 the 67th deployed to Al Kharj, Saudi Arabia in support the Joint Task Force-Southwest Asia. During this operation SOUTHERN WATCH deployment, the squadron flew over 1100 sorties and 3700 hours while enforcing United Nations resolutions and sanctions against Iraq.

F-15C 78-0521 is seen here at Kadena AB in March 1995 marked as the 67th Fighter Squadron (FS) Flagship and carrying the red tail stripe of the 67th FS. (Alec Fushi)

F-15C 78-0501 is seen here at Kadena AB in December 1995 carrying the red tail stripe of the 67th FS. (Alec Fushi)

F-15C 78-0522 of the 67th FS is seen here at Kadena AB in December 1995. (Alec Fushi)

F-15D 78-0571 is seen here at Kadena AB in December 1995 carrying 67th FS markings. (Alec Fushi)

UNITED STATES AIR FORCES EUROPE UNITS

BITBURG AIR BASE, GERMANY, 36th FIGHTER WING

22nd and 53rd Fighter Squadrons, 525th Tactical Fighter Squadron (525th TFS was inactivated in 1992)
Received F-15A/B in 1977 – Later re-equipped with F-15C/D.

The USAFE's 36th TFW was the first European user of the F-15, reaching its full strength of F-15A/B aircraft on September 30, 1977. The Wing was assigned the air defense of NATO's central front and expected to be the first unit in combat should a European war break out. The Wing converted to F-15C/Ds in 1980-81; the first F-15C arriving at Bitburg on August 22, 1980, with the last F-15A departing on November 14, 1981. Approximately 12 F-15C/D aircraft were transferred from Wing inventory to Royal Saudi Air Force (RSAF) during September 1990. This was to reinforce RSAF in response to Iraq's invasion of Kuwait. Shortly thereafter, virtually the entire Wing deployed to Southwest Asia for combat operations. Units of the 36th FW returned to Bitburg following Desert Storm. The 36th TFW was redesignated as the 36th Fighter Wing on October 1, 1991. On September 16, 1993, the Wing stood down from its "Zulu Alert" air defense alert posture. On October 1, 1993, in conjunction with the closure of Bitburg AB, the 36th began reassigning its F-15s to other bases. The 36th FW was inactivated on October 1, 1994. The 36th FW used BT tail codes in reference to Bitburg AB, where the Wing was based.

(Boeing via Marty Isham)

Right: (Don Logan Collection)
Below: (Boeing)

F-16C 79-0036 is seen marked as the 36th TFW Flagship in June 1985. (Don Logan Collection)

F-16C 79-0036 is seen marked as the 36th TFW Flagship on July 25 1986, with a colorful red white and blue nose stripe and rudder markings added. (Craig Kaston Collection)

F-15C 84-0005 is seen in September 1990 with a multi-color tail stripe. (Paul Hart Collection)

F-15C 80-0020 is seen in June 1982 carrying the tail markings for William Tell 82. (Don Logan Collection)

F-15C 79-0057 is seen here at Langley AFB on April 1, 1993 with a multi-colored tail stripe. (Brian C. Rogers)

22nd FIGHTER SQUADRON
STINGERS

The 22nd TFS received its F-15s, in 1977, replacing F-4Es. The 22nd did not deploy aircraft during Operation Desert Storm, although both pilots and maintenance personnel flew with the other two squadrons during the conflict. On October 1, 1991 the 22nd TFS was redesignated as the 22nd FS. The 36th FW's last three F-15s (from the 22nd FS) departed Bitburg for the United States on March 18, 1994. The squadron was inactivated on April 1, 1994. The 22nd FS used BT tail codes and wore a red tail stripe.

F-15C 84-0022 is seen marked as the 22nd Fighter Squadron (FS) Flagship and carrying the red tail stripe of the 22nd FS. (Paul Hart Collection)

F-15C 84-0022 is seen here at Langley AFB on April 1, 1993 marked as the 22nd AMU Flagship and carrying the red tail stripe of the 22nd FS. (Brian C. Rogers)

F-15A 76-0014 is seen here at RAF Alconbury on March 13, 1978 carrying the red tail stripe of the 22nd TFS. (Michael France)

F-15A 75-0056 is also seen at RAF Alconbury on March 13, 1978 in 22nd TFS markings. (Michael France)

F-15C 79-0060 is seen here at Luke AFB on March 13, 1978 carrying 22nd TFS markings. (Ben Knowles)

F-15C 80-0006 is seen on April 18, 1984 carrying the red tail stripe of the 22nd TFS. (Scott Wilson)

F-15C 80-0006 is seen here at Langley AFB on April 1, 1993 in 22nd FS markings. (Brian C. Rogers)

F-15C 79-0078 is seen here at Langley AFB on April 1, 1993 carrying the red tail stripe of the 22nd FS. (Brian C. Rogers)

258

53rd FIGHTER SQUADRON
NATO TIGERS

The 53rd TFS received its F-15s, replacing F-4Es in 1977. The 53rd deployed 24 F-15Cs to Prince Sultan AB, Al-Kharj, Saudi Arabia on December 20, 1990 as part of Operation Desert Shield. The squadron scored 11 air-to-air kills during and immediately after the Gulf War. On October 1, 1991 the 53rd TFS was redesignated as the 53rd FS. On February 25, 1994, the 53rd FS moved with its F-15s and personnel to Spangdahlem AB, Germany, joining the 52nd FW. The 22nd FS used BT tail codes and officially wore a yellow stripe on the vertical stabilizer, but its aircraft generally flew with yellow and black tiger stripes to celebrate the squadron's name. The 53rd periodically hosted the NATO Tiger meet, and usually painted one aircraft in tiger motif to participate.

Below: F-15C 84-0001 is seen here at Langley AFB marked as the 53rd Fighter Squadron (FS) Flagship and carrying the yellow and black tiger tail stripe of the 53rd FS. (Brian C. Rogers)

Bottom: F-15C 84-0001 is seen here at Langley AFB marked as the 53rd Tactical Fighter Squadron (FS) Flagship and carrying the yellow and black tiger tail stripe of the 53rd TFS. (Don Logan Collection)

F-15C 79-0053 is seen marked as the 53rd TFS Flagship and carrying the yellow and black tiger tail stripe of the 53rd TFS. (Kevin Patrick Collection)

Above: F-15C 84-0053 is seen marked as the 53rd Fighter Squadron (FS) Flagship and carrying special yellow and black tiger tail markings. (Don Logan Collection)

Right: F-15C 84-0021 is seen marked as the 53rd AMU Flagship and carrying the black tiger stripes on the tail and fuselage. (Don Logan Collection)

F-15B 75-0088 is seen landing at Greenham Common on June 26, 1978. The aircraft carries a tiger head on the nose. (Michael France)

F-15A 76-0009 is seen on September 23, 1978 carrying the yellow and black tiger tail stripe of the 53rd TFS. (Michael France)

F-15C 84-0021 is seen here at Langley AFB on April 4, 1993 carrying the yellow and black tiger tail stripe of the 53rd FS. (Brian C. Rogers)

525th TACTICAL FIGHTER SQUADRON
BULLDOGS

The first 23 operational F-15A/B aircraft for the 525th TFS arrived from Langley AFB on April 27, 1977. First to deploy to Desert Shield was the 525th, taking ten F-15Cs to Incirlik AB, Turkey on December 16, 1990. On January 17, 1991, the 525th deployed its remaining 14 F-15Cs to Incirlik. During Operation Desert Storm, the 525th TFS scored five air-to-air kills. The squadron was inactivated on March 31, 1992 as the situation in Europe improved. The 525th TFS used BT tail codes and wore a blue tail stripe.

F-15B 75-0087 is seen here at RAF Alconbury on May 5, 1980 carrying the blue tail stripe of the 525th Tactical Fighter Squadron. (Michael France)

F-15C 79-0048 is seen here at RAF Alconbury on June 2, 1982 marked with "Mad Dog" on the nose and carrying the blue tail stripe of the 525th Tactical Fighter Squadron. (Michael France)

F-15D 79-0006 is seen here at Luke AFB on November 8, 1981 carrying the blue tail stripe of the 525th Tactical Fighter Squadron. (Michael France)

F-15C 79-0044 is seen on August 6, 1983 carrying the blue tail stripe of the 525th TFS. (Scott Wilson)

F-15C 79-0073 is seen here at Luke AFB in June 1990 carrying the blue tail stripe of the 525th Tactical Fighter Squadron. (Ben Knowles)

RAF LAKENHEATH, GREAT BRITAIN, 48th FIGHTER WING

492nd (F-15E), 493rd (F-15C), and 494th FS (F-15E)

A late-comer to the F-15 was the 48th Fighter Wing, finally trading their General Dynamics F-111s for F-15Es beginning on February 21, 1992. The Wing has been based at Lakenheath since 1960, and is assigned to USAFE, 3rd Air Force. The 492nd FS achieved IOC with the F-15E on April 20, 1992. The 493rd FS began converting to the F-15C/D in December 1992. The 494th FS achieved IOC on August 13, 1992.

The 48th Fighter Wing is presently based at Royal Air Force Station Lakenheath, United Kingdom. The unit was activated on January 15, 1941 as the 48th Bombardment Group (Light) at Savannah Air Base, Georgia. Redesignated as the 48th Fighter Bomber Group, it deployed with its P-47s to the European Theater of operations. During April and May 1944, 48th P-47 Thunderbolts took part in the pre-invasion softening of the French coastal defenses just before the invasion of Normandy. The 48th Fighter Group inactivated at Seymour-Johnson Field, North Carolina on November 7, 1945.

On July 10, 1952 the 48th Fighter Bomber Wing was activated at Chaumont Air Base, France. The long-standing cooperation between the United States and France brought about a decision by the U.S. Air Force to give the 48th Fighter Bomber Wing the descriptive name, Statue of Liberty Wing. On July 4, 1954 the 48th was officially designated as the Liberty Wing by the U.S. Air Force, making the 48th the only U.S. Air Force unit with both a numerical and descriptive designation.

The 48th Fighter Bomber Wing was redesignated as the 48th Tactical Fighter Wing on July 8, 1958. On January 15, 1960 the 48th moved to RAF Lakenheath, United Kingdom. On January 7, 1972 the first F-4D Phantom II joined the Wing at Lakenheath replacing the Wing's F-100 aircraft. On March 1, 1977 as part of Operation Ready Switch, F-111Fs began replacing the Wing's F-4Ds. During April, 1986, 18 aircraft participated in Operation El Dorado Canyon(the U.S. attack against Libya). In support of Desert Shield the Wing deployed F-111Fs to Taif, Saudi Arabia, and by the beginning of Desert Storm 66 F-111Fs, had deployed. Combat operations began starting on January 17, 1991. During Desert Storm 48th aircrews flew 2,500 combat sorties into occupied Kuwait and Iraq, destroying or damaging 920 Iraqi tanks and armored person-

F-15E 90-0248 is seen marked as the 48th Fighter Wing Flagship (Joe Sadler)

nel carriers, 245 hardened aircraft shelters, 113 bunkers and 160 bridges with no aircraft lost due to enemy action.

On October 1, 1991, the 48th Tactical Fighter Wing was redesignated the 48th Fighter Wing. During February 1992 F-15Es joined the Wing, making the 48th the first in USAFE to operate the F-15E Strike Eagle. The F-15Es are assigned to the 492nd and 494th Fighter Squadrons. In January 1994 F-15Cs joined the Wing, assigned to the 493rd Fighter Squadron.

The Wing makes regular deployments to Aviano, Italy, to provide muscle for Operation 'Deny Flight' over Bosnia, and was involved in a UN- authorized strike against a Serbian airfield at Udbina, Croatia on November 21, 1994. A second strike was carried out a few days later against a surface-to-air missile site near Bihac. During August-September 1995 48th Wing aircraft flew strikes against Bosnia-Serb ground forces as part of Operation 'Deliberate Force', marking the first time F-15Es employed GBU-15 optically-guided 2,000 pound bombs against hostile targets. The Wing also provided aircraft to Operation Provide Comfort when it sent a detachment to Incirlik, Turkey.

Liberty Wing F-15Es deployed to Aviano Air Base, Italy, in February 1994 to participate in Operation Deny Flight. In November 1994 Liberty Wing F-15Es bombed Udbina Airfield in Croatia and a surface-to-air missile site near Bihac that posed a threat to NATO aircraft. In August and September 1995 Liberty Wing Strike Eagle crews struck Bosnia-Serb ground forces as part of NATO's Operation Deliberate Force. This strike was the first time F-15Es employed GBU-15 optically- guided 2,000 pound bombs against hostile targets.

Action in the Balkans continued with Operation Allied Force. Strike Eagles from the 494th FS were deployed to Aviano AB, Italy to fly strike missions into Kosovo and Serbia. Additional Strike Eagles launched from their home base of RAF Lakenheath to fly their strike missions. The 48th FW F-15Es flew over 1,000 sorties and 3,700 hours. They delivered over 2,770,000 pounds on ordnance (Laser Guided Bombs, GBU-15s and AGM-130s) against targets in both Kosovo and Serbia. F-15Cs of the 493rd FS deployed to Cervia AB, Italy flying combat air patrols. They scored four of the six allied aerial victories of Operation Allied Force, shooting down four Sebrian MiG-29s.

Above: F-15E 91-0313 is seen marked as the 3rd Air Force Flagship. The 3rd Air Force emblem can be seen on the CFT near the right intake. (Don Logan Collection)

Right: F-15E 91-0313 is seen marked as the 3rd Air Force Flagship. The three 48th Fighter Wing Squadron emblems can be seen on the CFT near the left intake. (Joe Sadler)

Right: F-15E 90-0248 is seen marked as the 48th Fighter Wing Flagship (Joe Sadler)

Below: F-15C 86-0166 is seen here at Langley AFB on January 20, 1995 marked as the 48th Operations Group (OG) Flagship. (Brian C. Rogers)

F-15E 90-0262 is seen here at Nellis AFB in April 1996 marked as the 48th Operations Group (OG) Flagship. (Alec Fushi)

F-15E 91-0604 is seen marked for the air show at RAF Waddington and the Tiger Meet in 1998. (Joe Sadler)

In preparation for the 48th Fighter Wing's change over from F-111Fs to F-15s, the 48th Logistics Support Squadron (LSS) received two F-15s to be used as maintenance trainers. F-15B 76-0124 was used to mimic the F-15E for training purposes, with F-15A 76-0029 used to mimic the F-15C. These two aircraft assisted in training over 220 48th Fighter Wing maintenance personnel. 76-0029 was written off and salvaged in 1997. 76-0124 seen here is still being used for training at RAF Lakenheath. (Joe Sadler)

F-15B 76-0124 is seen here at RAF Lakenheath in earlier 48th LSS tail markings (Joe Sadler)

492nd FIGHTER SQUADRON
BOWLERS or MADHATTERS

The 492nd Fighter Squadron Madhatters employ the F-15E Strike Eagle. The 492nd Fighter Squadron, originally constituted as the 55th Bombardment Squadron, was activated in January 1941, and redesignated as the 492nd Fighter Bomber Squadron in August 1943. The squadron flew A-18, A-20, A-35, P-39, P-40 and P-47 aircraft during World War II, serving with distinction. Its was inactivated in November 1945.

The squadron was reactivated at Chaumont AB, France, in July 1952. During this time, the squadron received F-84Gs (July 1952 - February 1954), F-86s (November 1953 - November 1956), and F-100Ds (September 1956 - April 1972). In 1960, the squadron transferred as part of the 48th TFW to RAF Lakenheath, England. They transitioned to F-4Ds in October 1971, and to F-111Fs in March 1977. In March 1992, who became USAFE's first unit to transition to the F-15E Strike Eagle. The squadron was declared mission-ready July 1, 1993.

The 492nd FS took part in Operation Allied Force by deploying six F-15Es to Aviano Air Base, Italy to augment their sister squadron the 494th EFS already in place at Aviano. The 492nd provided additional augmentation of Operation Allied Force by flying a limited number of combat Missions from their home base. On May 3, 1999, the 492nd Expeditionary Fighter Squadron launched their first combat missions in support of Operation Allied Force from RAF Lakenheath.

F-15E 90-0251 is seen here at RAF Lakenheath marked as the 492nd Fighter Squadron Flagship. (Joe Sadler)

F-15E 90-0260 is seen in November 1995 at Nellis AFB marked with the blue tail stripe of the 492nd Fighter Squadron. (David F. Brown)

F-15E 91-0303 is seen in April 1996 at Nellis AFB . (Alec Fushi)

F-15E 91-0312 is seen on October 23, 1996 marked with the blue tail stripe of the 492nd Fighter Squadron. (Pete Becker)

F-15E 91-0308 is seen in April 1996 at Nellis AFB. (Alec Fushi)

493rd FIGHTER SQUADRON
GRIM REAPERS

The 493rd Fighter Squadron is the air superiority unit of the Liberty Wing, flying MSIP F-15C Eagles. The 493rd was originally activated as 56th Bombardment Squadron (Light) on January 15, 1941, at Savannah, Georgia, flying the A-18. On May 23, 1941, operations moved to Will Rogers Field, Oklahoma, and crews began flying the A-20, and was reassigned to Savannah on February 27, 1942, and began conducting anti-submarine patrols. The 56th was redesignated as the 493rd Fighter Bomber Squadron in August 1943. The squadron saw action in World War II flying the P-47 Thunderbolt in the European Theater. The squadron was credited with 11 aerial victories by the end of the war. The squadron then moved to Seymour-Johnson AFB, North Carolina, and was inactivated November 7, 1945.

Reactivated June 25, 1952, the 493rd Fighter Bomber Squadron was assigned to Chaumont AB, France, flying the F-84G Thunderjet. In 1953, the squadron began flying the F-86 Sabre and in 1956 changed to the new F-100 Super Sabre. On July 8, 1958, the squadron was redesignated the 493rd Tactical Fighter Squadron. The squadron moved with the 48th TFW on January 6, 1960 to RAF Lakenheath, and continued to fly the F-100 Super Sabre until 1972. From 1972 to 1977, the 493rd flew the F-4D Phantom II until adopting the F-111F Aardvark. The 493rd Tactical Fighter Squadron received its present designation in February 1992, becoming the 493rd Fighter Squadron. As a result of the reassignment of the F-111F to Cannon AFB, New Mexico, the squadron was inactivated on December 14, 1992.

The 493rd Fighter Squadron was reactivated, flying air superiority F-15Cs on January 7, 1994, and was declared mission-ready June 12, 1994. It returned to Operation Provide Comfort from June to December 1994. Since September 1995, the Grim Reapers have participated in 12 deployments to four different countries, performing a variety of missions. These missions included air patrol missions over Northern Iraq to enforce the United Nations no fly zone in support of Operations Provide Comfort and Northern Watch, and participation in the Joint Chiefs of Staff exercise African Eagle 1996 at Morocco. In 1998 the 493rd returned to Incirlik AB, Turkey for 120 days supporting Operation Northern Watch.

The 493rd FS, as part of Operation Allied Force deployed to Cervia AB, Italy flying combat Air Patrol missions over the Balkans. The 493rd Expeditionary Fighter Squadron deployed 18 F-15C/Ds from RAF Lakenheath during February 1999 and were prepositioned at Cervia AB, Italy as part of the 501st Expeditionary Operations Group. They began flying combat on March 24, the first night of Operation Allied Force strikes, shooting down 2 Yugoslav MiG-29s. Two more MiG-29s were shot down on March 26. All the MiGs were shot down using the AIM-120 AMRAMM. The 493rd EFS was augmented by pilots and ground personnel from the 54th FS stationed at Elmendorf AFB, Alaska. The 54th FS deployed six F-15C to RAF Lakenheath, The aircraft remained at Lakenheath and were used for training by the 492nd pilots not deployed. By May 20 the 493rd EFS had flown 1,000 sorties. When the bombing halt occurred on June 10, the 493rd EFS had accumulated over 1,100 sorties and 5,500 flight hours. The 18 493rd F-15Cs returned to RAF Lakenheath on June 25 and 26, 1999.

F-15C 86-0164 is seen here at RAF Lakenheath marked as the 493rd Fighter Squadron Flagship. (Joe Sadler)

F-15C 86-0173 is seen here at Tyndall AFB on October 23, 1996 marked with the black tail stripe of the 493rd Fighter Squadron. (Nate Leong)

F-15C 86-0169 is seen here at Nellis AFB in November 1995. 86-0169 scored an aerial victory during Operation Allied Force, shooting down a Serbian MiG-29 on March 24, 1999. (David F. Brown)

494th FIGHTER SQUADRON
PANTHERS

The 494th Fighter Squadron flies F-15E Strike Eagles. The 494th Fighter Squadron began as the 57th Bombardment Squadron (Light) activated January 20, 1941, flying the A-18 trainer and A-20 Havoc aircraft. On August 10, 1943, the 57th was redesignated the 494th Fighter Bomber Squadron, flying the famous P-40 Warhawk and the P-39 Airacobra. In May 30, 1944, the 494th Fighter Bomber Squadron was assigned to Ibsley, England and redesignated the 494th Fighter Squadron, flying the P-47 Thunderbolt. After flying in numerous World War II campaigns in Europe, the 494th Fighter Squadron was inactivated November 7, 1945, at Seymour-Johnson Field, North Carolina.

On June 25, 1952, the 494th Fighter Squadron was reactivated and redesignated as the 494th Fighter Bomber Squadron. The Panthers transferred to Chaumont AB, France, assigned F-84 Thunder Streaks. While at Chaumont, the Panthers transitioned to the F-86 Sabre Jet, which they flew from 1954 to 1956, then to the F-100 Super Sabre. In 1958 the 494th Fighter Bomber Squadron was redesignated as the 494th Tactical Fighter Squadron. It was reassigned to RAF Lakenheath on January 15, 1960.

In 1972, the 494th Tactical Fighter Squadron Panthers received their first F-4D Phantom IIs as part of a project called Creek Phantom II. The 494th Tactical Fighter Squadron replaced the F-4s with the F-111F Aardvark in the spring of 1977. February 1, 1992, became the 494th Fighter Squadron. During 1992 the Panthers gave up the F-111F Aardvarks receiving F-15E Strike Eagles in their place.

Since the conversion to the F-15E aircraft, the Panthers have participated in Operation Provide Comfort over Northern Iraq during November 1993 through February 1995; Operation Deny Flight over Bosnia, June through October 1995; Operation Deliberate Force, the North Atlantic Treaty Organization's Bosnian bombing campaign, August through September 1995; Operations Joint Endeavor and Decisive Edge over Bosnia, January through March

1996; and Operation Northern Watch over Northern Iraq, January April 1997.

The 494th FS took part in Operation Allied Force flying combat missions from Aviano AB Italy On March 22 the 494th deployed F-15Es as part of the 494th Expeditionary Fighter Squadron to Aviano AB, Italy assigned to the 31st Air Expeditionary Wing. Six additional F-15Es from the 492nd FS were deployed to Aviano AB to augment the 494th aircraft. The 494th EFS continued to fly strike missions until the bombing halt on June 10, 1999. All 26 of the F-15Es deployed to Aviano AB returned to RAF Lakenheath on June 23 and 24, 1999.

F-15E 91-0314 is seen here at RAF Lakenheath marked as the 494th Fighter Squadron Flagship. (Joe Sadler)

F-15E 91-0314, the 494th Fighter Squadron Flagship, is seen here at RAF Lakenheath at the end of the runway receiving its final pre-takeoff checks. (Joe Sadler)

F-15E 91-0320 is seen here at Nellis AFB in April 1996 marked with the red tail stripe of the 494th Fighter Squadron. (Alec Fushi)

F-15E 91-0328 is seen here at Nellis AFB in April 1996. (Alec Fushi)

F-15E 91-0323 is seen in September 1994 in 494th Fighter Squadron markings. (David F. Brown)

494th Fighter Squadron F-15E 91-0601 is seen in October 1996. (Alec Fushi)

F-15E 91-0602 is seen here at Tyndall AFB on October 22, 1996 marked in 494th Fighter Squadron markings. (Nate Leong)

SOESTERBERG, THE NETHERLANDS, 32nd FIGHTER GROUP

32nd Fighter Squadron, Soesterberg, Netherlands.

32nd FIGHTER SQUADRON
WOLFHOUNDS

One of the more interesting squadrons in the Air Force was the 32nd FS based at Camp New Amsterdam, Soesterberg, The Netherlands. This squadron was under operational control of NATO through the Royal Netherlands Air Force, and their CR tail code was short for CRown, a theme also represented on their unit crest. All squadron aircraft had an orange stripe outlined in green on the vertical stabilizers, although this squadron (and most of the Air Materiel Command units) usually wore the stripe about a foot lower than other Air Force units. As part of Operation Coronet Sandpiper, the 32nd received its first F-15A/B on December 13, 1978, replacing F-4Es. The unit received its first F-15C on June 2, 1980, and completed conversion to the F-15C/D on November 19, 1980. The 32nd transferred half of its F-15C/D force to Royal Saudi Air Force in the fall of 1990. On January 17, 1991, the 32nd deployed five F-15Cs to Incirlik, Turkey and recorded one air-to-air victory over an Iraqi MiG-23. On November 30, 1991 the 32nd TFS was redesignated as the 32nd FS. The F-15C/Ds were replaced by MSIP (Multi-Stage Improvement Program) F-15A/Bs in 1992. The unit began phasing down operations for inactivation, with the first F-15s transferred out (to the 102nd FW, Massachusetts ANG) on October 1, 1993. In early 1994, the last F-15s of the 32nd Fighter Group left Soesterberg, ending nearly 40 years of USAF presence at this Dutch base. The unit was redesignated 32nd Air Operations Group, a non-flying support organization, on July 1, 1994.

Below: F-15A 77-0100 is seen on October 13, 1993 marked as the 32nd Fighter Squadron Flagship. (David F. Brown Collection)

F-15C 79-0015 is seen in a steep climb marked with the orange tail stripe of the 32nd Fighter Squadron. (Boeing)

F-15C 81-0049 is seen carrying special markings of the 32nd FS Wolfhounds on the nose. (USAF via Robert F. Dorr)

F-15C 79-0030 is seen on takeoff. (USAF via Robert F. Dorr)

F-15A 77-0100 is seen on October 13, 1993 marked as the 32nd Fighter Squadron Flagship. (Tom Kaminski Collection)

F-15A 77-0085 is seen in May 1984 in 32nd Fighter Squadron markings. (Jim Goodall)

F-15A 79-0018 is seen on October 13, 1993 marked as the 32nd Fighter Squadron Flagship. (Tom Kaminski Collection)

SPANGDAHLEM AIR BASE, GERMANY, 52nd FIGHTER WING

53rd Fighter Squadron F-15C/D

53rd FIGHTER SQUADRON
NATO TIGERS

When the 36th FW at Bitburg AB was closed down, one of its F-15C/D fighter squadrons was transferred to nearby Spangdahlem Air Base. The 53rd FS was moved to Spangdahlem on February 25, 1994, and became part of the 52nd Fighter Wing. At Spangdahlem AB it was equipped with 18 F-15C/D aircraft. The 52nd Wing is the largest fighter operation and most versatile Fighter Wing in the US Air Forces, Europe, with four fighter squadrons and an air control squadron. In addition to the 53rd FS, both the 22nd and 23rd Fighter Squadrons fly F-16s, while the 81st FS employs the A/OA-10. Since July 1993, the Wing has had F-15, A-10, and F-16 aircraft deployed to Aviano Air Base, Italy, in support of Operation Deny Flight. The F-15s left the 52nd FW in March 1999 when the 53rd FS was inactivated. Six of its 18 aircraft were sent to the 493rd FS at Lakenheath, with the remaining twelve aircraft returned to the 1st FW at Langley AFB in the United States.

F-15C 80-0052 is seen here at Nellis AFB in April 1996 marked as the 52nd Fighter Wing Flagship. (Alec Fushi)

F-15C 84-0001 is seen here at Nellis AFB in April 1996 marked as the 53rd Fighter Squadron Flagship. (Alec Fushi)

F-15C 84-0003 is seen here at Tyndall AFB in October 1994 marked with the yellow and black tiger tail stripe of the 53rd Fighter Squadron. (David F. Brown)

F-15C 80-0012 is seen taxiing at Nellis AFB in April 1996. (Alec Fushi)

F-15C 84-0019 is seen here at Nellis AFB in April 1996. Visible on the fuselage under the windshield are two Iraqi flags representing the two shoot-downs scored by this aircraft in Operation Desert Storm. (Alec Fushi)

F-15C 84-0025 is seen here at Tyndall AFB in October 1994 while attending William Tell 1994. (David F. Brown)

F-15C 84-0027 is seen in April 1996. Visible on the fuselage under the windshield are two Iraqi flags representing the two shoot-downs scored by this aircraft in Operation Desert Storm. (Alec Fushi)

F-15C 84-0010 is seen here at Langley AFB on October 12, 1994. Visible on the fuselage under the windshield is a single Iraqi flag representing the shoot-down scored by this aircraft in Operation Desert Storm. (Brian C. Rogers)

F-15C 84-0009 is seen here at Nellis AFB in April 1996 marked with the yellow and black tiger tail stripe of the 53rd Fighter Squadron. (Alec Fushi)

53rd Fighter Squadron F-15C 80-0004 is seen here at Nellis AFB in April 1996. (Alec Fushi)

AIR NATIONAL GUARD UNITS

The F-15 has become one of the mainstays of the ANG, with seven squadrons presently operating the F-15A/B.

FLORIDA AIR NATIONAL GUARD, JACKSONVILLE INTERNATIONAL AIRPORT
125th FIGHTER WING

159th FIGHTER SQUADRON
JACKSONVILLE JAGUARS
The 159th FS from Florida's 125th FW began receiving F-15A/Bs in June 1995. The Wing's F-15 sit alert at Homestead AFB, Florida.

Below: F-15B 76-0125 is seen here at Jacksonville International Airport on August 12, 1998 marked as the 125th Fighter Wing Flagship. (Jerry Geer) Left: (Don Logan)

F-15B 76-0125 is seen here at Jacksonville International Airport on October 25, 1995. The jaguar in these early 125th Fighter Wing Flagship markings is on the nose, with the 125th FW lightning bolt on the tail. (Nate Leong)

F-15B 76-0125 is seen here at Hector Field, Fargo, North Dakota on August 20, 1999 marked as the 125th Fighter Wing Flagship. (Don Logan)

F-15A 75-0039 is seen on the ramp at Jacksonville International Airport on May 6, 1998 with the 125th FW lightning bolt on the tail. (Nate Leong)

F-15A 75-0078 is seen taxiing at Jacksonville International Airport on May 6, 1998. (Nate Leong)

GEORGIA AIR NATIONAL GUARD, DOBBINS ARB
116th FIGHTER WING

128th FIGHTER SQUADRON
The 128th FS belongs to the 116th FW, Georgia ANG. The squadron was based at Dobbins ARB, just outside Marietta, Georgia, and began receiving F-15s in March 1986. Originally the unit carried a broad dark stripe with GEORGIA on the vertical stabilizer, but began using GA tail codes before it replaced its F-15s with B-1Bs in 1996.

F-15A 74-0116 is seen in July 1987 marked as the 116th Tactical Fighter Wing Flagship. (David F. Brown)

F-15A 74-0128 is seen in July 1987 marked as the 128th Fighter Squadron Flagship. (David F. Brown)

Above: F-15B 73-0108 (TAC-1 the first F-15 delivered to Tactical Air Command) is seen here at the London, Ontario Air Show on June 2, 1991. (David F. Brown)

Right: F-15A 75-0068 is seen on takeoff at the London, Ontario Air Show on June 2, 1991. (David F. Brown)

Below: F-15A 75-0065 is seen in April 1991 with the low visibility tail stripe of the Georgia Air National Guard. (David F. Brown)

F-15A 75-0068 is seen in June 1989 with the full color tail stripe of the Georgia Air National Guard. (David F. Brown)

Above: F-15B 75-0088 is seen here at the London, Ontario Air Show on June 7, 1993 carrying the GA tail code adopted by the 116th Fighter Wing. The GA tail code became available following the inactivation of the 35th Fighter Wing at George AFB in June 1992. (David F. Brown)

Right: F-15A 75-0024 is seen in June 1993 with the GA tail codes of the 116th FW. (David F. Brown)

HAWAII AIR NATIONAL GUARD, HICKAM AFB, HONOLULU
154th FIGHTER WING

199th FIGHTER SQUADRON
THE KUKAIUMOKU

Based at Hickam AFB, the Hawaii ANG operates the 199th FS from the 154th FW. The unit was one of the last users of the F-4, replacing them with F-15A/Bs in the summer of 1987. The majority of the original group of aircraft came from the 43rd TFS, 21st TFW at Elmendorf, as that unit transitioned from the F-15A/B to the F-15C/D. The original group of aircraft were exchanged for F-15A/B MSIP aircraft from Holloman in 1992.

199th Fighter Squadron

F-15A 74-0098 is seen in May 1988 at Hickam AFB with the full color, high visibility markings of the Hawaiian Air National Guard. (Alec Fushi)

F-15A 77-0079 is seen on June 24, 1996 during a training mission. (Robert F. Dorr)

F-15A 77-0062 is seen on June 24, 1996 receiving fuel from KC-135R 60-0323 assigned to the 203rd Air Refueling Squadron, Hawaii Air National Guard. (Robert F. Dorr)

F-15A 74-0109 is seen in January 1992 at Hickam AFB with the 199th Fighter Squadron full color, high visibility markings. (Jerry Geer)

F-15A 74-0107 is seen on November 19, 1987 at Hickam AFB with Day-Glo speed brake and centerline tank. The high visibility Day-Glo paint was used to make identification of the aircraft easier during air combat training. (Wally Van Winkle)

F-15B 75-0081 is seen on November 22, 1987 at Hickam AFB with the black, high visibility markings of the Hawaiian Air National Guard. The painted over 22 tail code of the 18th Wing is visible on the tail. (Norris Graser)

F-15B 74-0140 is seen in June 1994 at Luke AFB with the non-standard camouflage and low visibility markings. (Kevin Patrick)

F-15A 77-0077 is seen in May 1988 at Hickam AFB with the black, high visibility markings. (Nate Leong)

F-15A 76-0114 is seen in May 1994 at Luke AFB with the 199th FS low visibility markings. (Kevin Patrick)

LOUISIANA AIR NATIONAL GUARD, NEW ORLEANS JOINT RESERVE BASE
159th FIGHTER WING

122nd FIGHTER SQUADRON
BAYOU MILITIA

The 122nd FS from Louisiana's 159th FW wear the tail code 'JZ,' short for 'jazz' in tribute to their New Orleans home. The 122nd was the first ANG squadron to receive F-15s, beginning in June 1985. The first aircraft came mostly from Luke, early FY73 models, with the original speedbrake with external stiffener, and a coat of Air Superiority Blue underneath the Compass Ghost gray. In 1991, the FY73 aircraft were traded for FY77 models, mostly from Holloman. In 1993 the squadron began receiving aircraft that had been processed through the MSIP. The squadron is now known as the Bayou Militia; the term Coonass Militia being deemed politically incorrect (a Coonass being a native of Louisiana). The squadron has deployed to Keflavik; Alborg, Denmark; Howard AFB, Panama; CFB Cold Lake, Canada; Incirlik, Turkey; and Bodo, Norway. The purple, green and gold stripes on the tails are the colors of Mardi Gras, and each flight of aircraft is designated by one of the colors.

F-15A 77-0067 is seen here at New Orleans Joint Reserve Base on August 16, 1995 marked as the 159th Fighter Group Flagship. (Don Logan)

F-15A 77-0120 is seen on May 26, 1993 marked as the 159th Consolidated Aircraft Maintenance (CAM) Squadron Flagship. The aircraft crashed shortly after takeoff at New Orleans Joint Reserve Base on June 8, 1998. (Nate Leong)

F-15A 77-0122 is seen on June 26, 1995 marked as the 122nd Fighter Squadron Flagship. (Nate Leong)

F-15A 77-0122 is seen in November 1986 marked as the 122nd Tactical Fighter Squadron Flagship. (Don Spering/AIR)

F-15A 73-0105 is seen on September 2, 1989 at Luke AFB with the tail stripe of yellow flight. (Douglas Slowiak/Vortex Photo Graphics)

Above: F-15A 73-0085 is seen on January 7, 1991 at NAS Glenview with the tail stripe of red flight. (Norris Graser)

Right: F-15A 73-0098 is seen on March 30, 1991 at NAS Glenview. (Norris Graser)

F-15A 77-0116 is seen on May 26, 1993 at Volk Field, Wisconsin. (Nate Leong)

F-15A 74-0125 is seen on May 11, 1991 at Luke AFB. (Douglas Slowiak/Vortex Photo Graphics)

F-15A 77-0134 is seen ready for takeoff. (David F. Brown)

F-15A 77-0114 is seen here at New Orleans Joint Reserve Base on August 16, 1995 marked with the multi-colored tail stripe of 159th Fighter Wing. (Don Logan)

MASSACHUSETTS AIR NATIONAL GUARD, OTIS AIR NATIONAL GUARD BASE
102nd FIGHTER Wing

101st FIGHTER SQUADRON

The 101st Fighter Squadron (previously a Fighter Interceptor Squadron) of the 102nd Fighter Wing, Massachusetts ANG, began receiving F-15s in September 1987. Most of the unit's aircraft originally came from the inactivating 5th FIS at Minot AFB. These aircraft were replaced, beginning in October 1993, by aircraft from the 32nd FS which had been based at Camp New Amsterdam, The Netherlands. Massachusetts Guardsmen have on several occasions intercepted and escorted Soviet long-range reconnaissance aircraft off the Atlantic coast of the US. The unit is based at Otis ANGB on Cape Code. An ANG Minuteman is superimposed over a map of the state on the vertical stabilizer of their aircraft, and the centerline tanks carries a 'Cape Cod' inscription superimposed on a harpoon.

F-15A 77-0102 is seen here at New Orleans Joint Reserve Base in December 1998 marked as the 102nd Fighter Wing Flagship. (Paul Hambleton)

Below: F-15A 76-0015 is seen here at Pease AFB, New Hampshire on September 09, 1989 marked as the 102nd Fighter Interceptor Wing (FIW) Flagship. (Barry Roop)

F-15A 77-0115 is seen here in September 1999 with MA tail codes. The 102nd began adding MA tail codes to their aircraft in Summer 1999. (Jerry Geer)

Above: F-15A 76-0027 is seen in September 1990 at Otis Air National Guard Base, home of the 102nd Fighter Wing, Massachusetts Air National Guard. (Ben Knowles)

Right: F-15A 76-0016 is seen in July 1988 with the minuteman and Massachusetts map of the 102nd Fighter Wing, Massachusetts Air National Guard. (Dave Brown Collection)

F-15A 77-0097 is seen on April 17, 1988 at Volk Field, Wisconsin. Although the aircraft has the minuteman and Massachusetts map of the 102nd Fighter Wing, Massachusetts Air National Guard, the 32nd Fighter Squadron emblem on the intake indicates its recent transfer from Soesterberg, The Netherlands. (Nate Leong)

F-15A 76-0039 is seen in September 1990 at Otis Air National Guard Base. (Ben Knowles)

F-15A 77-0113 is seen on August 24, 1994 at Volk Field, Wisconsin taxiing for a training mission. (Nate Leong)

MISSOURI AIR NATIONAL GUARD, LAMBERT FIELD
131st FIGHTER Wing

110th FIGHTER SQUADRON
LINDBERGH'S OWN

The 110th FS is assigned to the 131st FW, Missouri ANG, at the Lambert-St Louis Airport. The SL tail codes reflects their St Louis home. The squadron is located directly across the runway from the McDonnell Douglas factory that builds F-15s. The unit is assigned battlefield air superiority, and also provides dissimilar air combat training for ANG F-16 units. The 110th began converting from F-4Es in May 1991, and achieved Initial Operational Capability on 15th September 1991.

F-15A 77-0131 is seen here at Lambert Field, St Louis, Missouri on December 8, 1992 marked as the 131st Fighter Wing Flagship. In the background is the McDonnell Douglas factory where all F-15s were produced. (Nate Leong)

The 131st Fighter Wing Flagship leads a four-ship past the St Louis Arch and downtown St. Louis Missouri. (McDonnell Douglas via Robert F. Dorr)

F-15A 77-0136 is seen in September 1994 at Lambert Field, St Louis, Missouri. (Alec Fushi)

F-15A 77-0030 marked as the 131st Fighter Wing Flagship is seen taxiing at Lambert Field, St Louis on October 1, 1992. It carries the nose art "Pride of St. Louis". (Both Douglas Slowiak/Vortex Photo Graphics)

F-15A 76-0045 is seen on October 1, 1992 at Lambert Field, St Louis with the St. Louis Arch tail markings and the full color 110th FS emblem on the intake. (Douglas Slowiak/Vortex Photo Graphics)

F-15A 76-0033, seen here on October 1, 1992 carries the low visibility tail markings. (Douglas Slowiak/Vortex Photo Graphics)

F-15A 77-0117 is seen on February 24, 1995 at Lambert Field, St Louis, Missouri. The aircraft crashed on August 19, 1999 during a normal training mission. The pilot ejected safely. (Nate Leong)

In this photo taken on July 25, 1994, the Air National Guard emblem is visible on the inside of F-15A 77-0145's vertical stabilizer. (Norris Graser)

SPECIAL OK (OKLAHOMA ANG MARKINGS) - Never assigned to Oklahoma ANG.

Although F-15s were never assigned to the Oklahoma Air National Guard, F-15A 76-0045 is seen on August 10, 1991 at Lambert Field, St Louis with the OK tail codes of the 125th FS, Oklahoma ANG. (Above - Norris Graser, Below - Nate Leong)

OREGON AIR NATIONAL GUARD, PORTLAND INTERNATIONAL AIRPORT
142nd FIGHTER Wing

123rd FIGHTER SQUADRON
RED HAWKS

Oregon ANG's 123rd FS is part of the 142nd FW, and is based at the Portland International Airport. The unit received its first F-15 on 1st October 1989, with most of the F-15s coming from the disbanded 318th FIS at McChord AFB. Interestingly, the unit's assigned mission is to provide air defense of the U.S. west coast, the same role previously assigned to the 318th.

Left: (Craig Kaston)

F-15B 76-0142 is seen on August 20, 1994 marked as the 142nd Fighter Group Flagship. It carries the name "Portland City of Roses" on the inside of the vertical stabilizer. (Alec Fushi)

F-15A 77-0119 is seen on landing at Kingsley Field, Oregon on August 5, 1999. (Nate Leong)

F-15A 75-0061 is seen here at Tyndall AFB on October 23, 1996. It carries the 123rd FS red tail stripe and dark gray hawk on the tail. (Nate Leong)

F-15A 76-0106 is seen on August 10, 1993. Its tail markings, the 123rd FS hawk and serial number, are in low visibility light gray. (Jerry Geer)

F-15A 76-0098 is seen in December 1992. It carries the 123rd FS hawk in black on the tail. (Ben Knowles)

OREGON AIR NATIONAL GUARD, KINGSLEY FIELD, KLAMATH FALLS
173rd FIGHTER Wing

114th FIGHTER SQUADRON

The 173rd FW, based at Kingsley Field near Klamath Falls, recently transitioned from the F-16 to become an F-15 training Wing. Each ANG unit gave up a 'family model' (F-15B) so that the new training squadron could have a large number of two-seat F-15s. With the arrival of the new F-22 Raptor, the 325th FW at Tyndall will begin transitioning to the F-22, and Klamath Falls will begin training all F-15 pilots.

In 1982 the Air Force announced a proposal to establish an air defense schoolhouse for F-4 Phantom II aircrews under the Air National Guard. In mid-November 1988, the last F-4 class graduated. The first F-16 aircraft arrived at Kingsley Field in August 1988. The first Air Defense Fighter modified F-16 aircraft was received March 1, 1989 followed by the first F-16 student class on July 13, 1989. On June 27, 1996, the flying unit was officially redesignated the 173rd Fighter Wing. The 114th Fighter Squadron was retained as the flying component under the Wing's Operations Group.

Beginning in November, 1997, the Wing began plans to change aircraft from the F-16 to the F-15 Eagle. The unit remained as a schoolhouse helping to relieve shortages in F-15 training quotas for the total force. The last F-16s were flown out on February 26, 1998 to Tucson, Arizona. After extensive Maintenance and pilot training, F-15 flying operations began in June 1998.

Below: F-15A 75-0052 is seen taxiing at Kingsley Field, Klamath Falls, Oregon, the home of the 173rd Fighter Wing. (Nate Leong)

(Nate Leong)

F-15B 76-0127 is seen on the ramp at Kingsley Field, Klamath Falls, Oregon. (Nate Leong)

F-15A 75-0052 and F-15B 77-0165 are seen here on a training mission over northern California's Mount Shasta. (USAF)

F-15B 76-0139 is seen taxiing at Kingsley Field, Klamath Falls, Oregon. (Nate Leong)

CHAPTER SIX

F-15 Eagle with the USAF in Desert Storm

On August 1, 1990, Iraqi forces invaded Kuwait, on August 6, the US launched Operation Desert Shield to defend against any Iraqi moves southward against Saudi Arabia. On the same day the 1st Tactical Fighter Wing based at Langley AFB began deployment of its F-15C/Ds to Dhahran, Saudi Arabia. This was followed on August 12, with F-15Es from the only operational F-15E squadron, the 336th TFS, 4th TFW based at Seymour-Johnson AFB. The F-15Es arrived at Thumrait, Oman, and subsequently moved to Al Karj, Saudi Arabia.

The F-15C/Ds began to fly combat air patrols along with Saudi F-15Cs and British and Saudi Tornado F.Mk 3s. The F-15Es began training for strike missions. During one of the training missions, F-15E 87-0203 crashed on September 30, 1990, killing both crewmen.

A second round of Desert Shield deployments took place in November 1990. The 58th TFS, 33rd TFW deployed F-15C Eagles to Tabuk in western Saudi Arabia. The 53rd TFS, 36th TFW based at Bitburg in Germany also deployed to Tabuk. Aircraft of the 525th TFS, also from the 36th TFW joined the 7440th Composite Wing at Incirlik in Turkey. The 32nd TFS based at Soesterberg in the Netherlands also deployed to Incirlik. A second 4th TFW F-15E squadron, the 335th TFS moved to Al Kharj.

Operation Desert Storm began just after midnight on January 17, 1991. Most of the air-to-air engagements during the war were fought by the F-15C, and most of these by pilots of the 58th TFS. Thirty-four enemy aircraft (32 airplanes and 2 helicopters) were destroyed by USAF F-15Cs during the Gulf War, against zero losses. In addition, an F-15E (89-0487) shot down a H.500 Helicopter on February 11, 1991. Many of the kills were against Iraqi aircraft caught by chance or attempting to flee to Iran. There was very little of the dog fighting for which the F-15 had been built, instead most of the kills were made beyond visual range (BVR) by AIM-7 Sparrow missiles. Nine kills were made by the F-15C with the AIM-9 Sidewinder missile, and one kill was credited to a F-15C pilot who maneuvered his MiG-29 opponent into flying his aircraft into the ground. None of the enemy aircraft ever got close enough to be in range of the F-15C's 20-mm cannon.

F-15C 85-0102 was credited with three aerial victories during Desert Storm, On January 29, 1991 Captain David Rose piloting 85-0102 shot down a MiG-23, followed by Captain Anthony Murphy shooting down two Su-22s on February 7. Two F-15C pilots, Captain Thomas Dietz and Lt. Robert Hehemann, are credited with three aerial victories apiece, although one of each pilot's victories occurred on March 22, 1991, after the war was officially over.

Although the F-15E Strike Eagle was still not fully combat-ready, 48 F-15Es flew in the Gulf War. Most of their sorties were flown at medium altitudes, with the F-15E not getting a chance to demonstrate its low-level capabilities. Although by the end of the war only some of the F-15Es were equipped with LANTIRN targeting pods, pilots claimed that 80 percent of the laser-guided bombs dropped by F-15Es hit their targets. Because of the newness of the LANTIRN pod, difficulties were still being encountered in fully integrating the system on the F-15E. As a result, its full capabilities were not available at that time.

No F-15C/D Eagles were lost in combat, although two F-15E Strike Eagles were shot down by ground fire, one on January 18 (88-1689) and the other on January 19 (88-1692). The crew of the first plane was killed in the crash, the crew of the second was taken prisoner and later released.

After the war was officially over, F-15Cs continued to carry out combat air patrols, enforcing the "no-fly" restrictions on Iraqi fixed-Wing aircraft imposed under the terms of the cease-fire. On March 20, F-15C 84-0014 flown by Capt. John T. Doneski of the 22nd TFS shot down one of two Iraqi Su-22s with an AIM-9 missile, the other Su-22 making a hasty landing. On March 22, F-15C 84-0010 flown by Capt. Thomas N. Dietz of the 53rd TFS shot down another Su-22 violating the no-fly order. This was Capt. Dietz's third kill, having downed a pair of MiG-21s on February 6. The pilot of another F-15C, Lt. Robert Hehemann was able to claim a Pilatus PC-9 trainer which was flying in close vicinity of the downed Su-22 when its pilot baled out without a shot being fired. This was kill number three for Lt. Hehemann as well.

Above: (USAF)

Right: F-15D 82-0046 from the 27th FS, 1st Fighter Wing based at Langely AFB,
Virginia is seen during Operation Desert Shield on the ramp at Dhahran Air Base,
Saudi Arabia. (McDonnell Douglas via Robert F. Dorr)

F-15E wartime experience was handed over to the F-15 Combined Test Force (CTF) at Edwards AFB, which was still accomplishing developmental testing on the F-15E engine, software, radar, weapons, and LANTIRN systems. Additional testing still had to be completed for full set of F-15E weapons, including the Mk 20 Rockeye and CBU-87 cluster bombs, Mk-82 500 pound and Mk-84 1000 pound bombs, AGM-65 Maverick missiles, and GBU-10 and GBU-15 laser-guided bombs.

On April 14, 1994, there was a tragic "friendly fire" incident over northern Iraq, when a pair of F-15Cs of the 52nd Fighter Wing enforcing the "no-fly" rule mistakenly shot down two UH-60 Blackhawk helicopters, killing 26 American and United Nations personnel carrying out humanitarian aid to Kurdish areas of Iraq. One of the helicopters was destroyed by an AIM-120 AMRAAM, the other by a AIM-9 Sidewinder.

DESERT STORM UNITS

4th TACTICAL FIGHTER WING (PROVISIONAL)

When Operation 'Desert Shield' began on 6th August 1990, the 336th FS, 4th Wing, the only operational F-15E squadron in the Air Force, deployed arriving at Thumrait, Oman on 12th August 1990. The aircraft were subsequently moved to Al Kharj, Saudi Arabia when the 4th TFW(P) was established under the 14th Air Division (Provisional). The 335th FS F-15Es also deployed, and by the time

Operation Desert Storm began on January 17, 1991 there were 48 F-15Es in Saudi Arabia. In addition to the two F-15E squadrons, 4th TFW(P) also controlled the F-15Cs from the 53rd TFS, 36th TFW at Bitburg, and two Air National Guard F-16 squadrons (138th TFS/New York, and 157th TFS/South Carolina). The 336th returned home shortly after the end of hostilities, and the 335th a few of months later.

4404th WING (PROVISIONAL)

The 4404th Wing (Provisional) was activated on March 13, 1991, at Al Kharj Air Base. The original assets of the 4404th TFW(P) came from the 4th TFW(P), which had operated during the Gulf War. In June 1992 the Wing relocated to King Abdul Aziz Air Base, Dhahran,

F-15C 83-0026 from the 71st FS, 1st Fighter Wing based at Langely AFB, Virginia is seen on October 12, 1990 on the ramp at Dhahran Air Base, Saudi Arabia. (Robert F. Dorr)

where it was officially reactivated as the 4404th Wing (Provisional) on August 2, 1991.

On August 27, 1992, the United Nations restricted the Iraqi flight operations south of the 32nd parallel in response to Iraqi attacks on Shiite minorities in southern Iraq. Coalition forces from the US Navy, the British Royal Air Force, and the French Air Force joined the Wing in Operation 'Southern Watch', monitoring Iraqi compliance with the UN mandate. At the beginning of 1998, the Coalition continued to fly Southern Watch missions around the clock. Beyond the 'Southern Watch' mission, the Wing also supported Operation Restore Hope in December 1992 and evacuated over 600 US and foreign citizens from Yemen in May 1994. US Air Force aircraft, particularly F-15E squadrons, deployed to the 4404th as required to support operational needs.

The 4404th Wing did not remain at Dhahran. After a terrorist truck bomb killed 19 airmen, the Wing was ordered to move to a safer location within the Kingdom of Saudi Arabia. Prince Sultan Air Base was chosen as the relocation site , which meant that the 4404th was returning home to Al Kharj where it first originated. The morning of 5th August, the first of five C-5s landed and over a 45 day period, the Wing moved aircraft, personnel, and equipment from Riyadh and Dhahran to its new home at Prince Sultan Air Base.

The 4404th Wing employed 16 different types of aircraft, including the F-15C, F-15E, F-16, KC-135, KC-10, RC-135, E-3, U-2, C-130, HC-130, and C-21 aircraft which provide fighter, electronic combat, reconnaissance, command and control, air refueling, search and rescue, and cargo/troop transport capabilities. The 4404th Wing (Provisional) has over 4,500 personnel assigned to six geographically separated locations, in three countries in the Southwest Asia area of responsibility. The 4404th Wing was replaced by the 363rd Air Expeditionary Wing (AEW) on December 1, 1998.

DESERT STORM USAF F-15 AERIAL VICTORIES

DATE	TYPE	SERIAL #/TC	CREW	TYPE DEST	WEAPON	WING/SQDN
01-17-91	F-15C	85-0125/EG	Kelk	MiG-29	AIM-7	33TFW/58TFS
01-17-91	F-15C	85-0105/EG	Graeter	Mirage F.1	AIM-7	33TFW/58TFS
01-17-91	F-15C	85-0105/EG	Graeter	Mirage F.1	AIM-7	33TFW/58TFS
01-17-91	F-15C	83-0017/FF	Tate	Mirage F.1	AIM-7	1TFW/71TFS
01-17-91	F-15C	85-0108/EG	Draeger	MiG-29	AIM-7	33TFW/59TFS
01-17-91	F-15C	85-0107/EG	Magill (USMC)	MiG-29	AIM-7	33TFW/58TFS
01-19-91	F-15C	85-0122/EG	Underhill	MiG-29	AIM-7	33TFW/58TFS
01-19-91	F-15C	85-0114/EG	Rodriguez	MiG-29	(ACM)	33TFW/58TFS
01-19-91	F-15C	85-0099/EG	Pitts	MiG-25	AIM-7	33TFW/58TFS
01-19-91	F-15C	85-0101/EG	Tollini	MiG-25	AIM-7	33TFW/58TFS
01-19-91	F-15C	79-0069/BT	Prather	Mirage F.1	AIM-7	36TFW/525TFS
01-19-91	F-15C	79-0021/CR	Sveden	Mirage F.1	AIM-7	36TFW/525TFS
01-26-91	F-15C	85-0119/EG	Draeger	MiG-23	AIM-7	33TFW/59TFS
01-26-91	F-15C	85-0104/EG	Schiavi	MiG-23	AIM-7	33TFW/58TFS
01-26-91	F-15C	85-0114/EG	Rodriguez	MiG-23	AIM-7	33TFW/58TFS
01-27-91	F-15C	84-0025/BT	Denney	MiG-23	AIM-9	36TFW/53TFS
01-27-91	F-15C	84-0025/BT	Denney	MiG-23	AIM-9	36TFW/53TFS
01-27-91	F-15C	84-0027/BT	Powell	MiG-23	AIM-7	36TFW/53TFS
01-27-91	F-15C	84-0027/BT	Powell	Mirage F.1	AIM-7	36TFW/53TFS
01-29-91	F-15C	79-0022/BT	Watrous	MiG-23	AIM-7	32TFG/32TFS
01-29-91	F-15C	85-0102/EG	Rose	MiG-23	AIM-7	33TFW/60TFS
02-02-91	F-15C	79-0064/BT	Masters	Il-76	AIM-7	36TFW/525TFS
02-06-91	F-15C	84-0019/BT	Hehemann	Su-25	AIM-9	36TFW/53TFS
02-06-91	F-15C	84-0019/BT	Hehemann	Su-25	AIM-9	36TFW/53TFS
02-06-91	F-15C	79-0078/BT	Dietz	MiG-21	AIM-9	36TFW/53TFS
02-06-91	F-15C	79-0078/BT	Dietz	MiG-21	AIM-9	36TFW/53TFS
02-07-91	F-15C	85-0102/EG	Murphy	Su-22	AIM-7	33TFW/58TFS
02-07-91	F-15C	85-0102/EG	Murphy	Su-22	AIM-7	33TFW/58TFS
02-07-91	F-15C	85-0124/EG	Parsons	Su-22	AIM-7	33TFW/58TFS
02-07-91	F-15C	80-0003/BT	May	Mil-24 Helicopter	AIM-7	36TFW/525TFS
02-11-91	F-15C	79-0048/BT	Dingee	Mil-8 Helicopter*	AIM-7	36TFW/525TFS
02-11-91	F-15C	80-0012/BT	McKenzie	Mil-8 Helicopter*	AIM-7	36TFW/525TFS
02-14-91	F-15E	89-0487/SJ	Bennett/Bake	H.500 Helicopter	LGB	4TFW/335TFS
03-20-91	F-15C	84-0014/BT	Doneski	Su-22	AIM-9	36TFW/53TFS
03-22-91	F-15C	84-0010/BT	Dietz	Su-22	AIM-9	36TFW/53TFS
03-22-91	F-15C	84-0015/BT	Hehemann	PC-9	(ACM)	36TFW/53TFS

Mil-8 02-11-91 shared credit

79-0021 JANUARY 19, 1991 Mirage F.1 CR Tail Code (Alec Fushi)

79-0022 JANUARY 29, 1991 MiG -23 BT Tail Code (Don Logan Collection)

79-0048 FEBRUARY 11, 1991 1/2 Mil-8 Helicopter BT Tail Code (Don Logan Collection)

Right: 79-0064 FEBRUARY 2, 1991 IL-76 BT Tail Code (Don Logan Collection)

Below: 79-0069 JANUARY 19, 1991 Mirage F.1 BT Tail Code (Don Logan Collection)

79-0078 FEBRUARY 6, 1991 Two MiG-21s BT Tail Code (Brian C. Rogers)

80-0003 FEBRUARY 7, 1991 Mil-24 Helicopter BT Tail Code (David F. Brown)

80-0012 FEBRUARY 11, 1991 1/2 Mil-8 Helicopter BT Tail Code (Ben Knowles)

80-0017 JANUARY 17, 1991 Mirage F.1 FF Tail Code (Jerry Geer)

84-0010 MARCH 22, 1991 Su-22 BT Tail Code (Jerry Geer Collection)

84-0014 MARCH 20, 1991 Su-22 BT Tail Code (Tom Kaminski Collection)

84-0015 MARCH 22, 1991 PC-9 BT Tail Code (Brian C. Rogers)

Right: 84-0019 FEBRUARY 6, 1991 Two Su-25 BT Tail Code (Tom Kaminski Collection)

Below: 84-0025 JANUARY 27, 1991 Two MiG-23 BT Tail Code (Nate Leong)

84-0027 JANUARY 19, 1991 MiG-23, Mirage F.1 BT Tail Code (Paul Hart Collection)

85-0099 JANUARY 19, 1991 MiG-25 EG Tail Code (Douglas Slowiak/Vortex Photo Graphics)

85-0101 JANUARY 29, 1991 MiG-25 EG Tail Code (Alec Fushi)

85-0102 JANUARY 29, 1991 MiG-23 EG Tail Code. FEBRUARY 7, 1991 Two Su-22s. (Ben Knowles)

85-0104 JANUARY 26, 1991 MiG-23 EG Tail Code (Alec Fushi)

85-0105 JANUARY 17, 1991 Two Mirage F.1s EG Tail Code (Alec Fushi)

85-0107 JANUARY 17, 1991 MiG-29 EG Tail Code (Alec Fushi)

85-0108 JANUARY 17, 1991 MiG-29 EG Tail Code (Ben Knowles)

85-0114 JANUARY 19, 1991 MiG-29 EG Tail Code. JANUARY 26, 1991 MiG-23. (Peter Becker/AIR)

85-0119 JANUARY 26, 1991 MiG-23 EG Tail Code (Jim Geer)

85-0122 JANUARY 19, 1991 MiG-29 EG Tail Code (Robert F. Dorr)

85-0124 FEBRUARY 7, 1991 Su-22 EG Tail Code (Alec Fushi)

85-0125 JANUARY 19, 1991 Mirage F.I EG Tail Code (Ben Knowles)

89-0487 (F-15E) FEBRUARY 14, 1991 Hughes 500 Helicopter EG Tail Code (Alec Fushi)

CHAPTER SEVEN

F-15 Eagle with the USAF in Operation Allied Force

On March 24, 1999 NATO forces launched Operation Allied Force to defend Kosovo against Serbian moves aimed at ethnic cleansing. This Operation was to be an air campaign designed to force the Serbian Government to cease their aggression against Kosovo by inflicting heavy losses against the Military and Political infrastructure of Serbia. F-15C/Ds from the 48th Fighter Wing were already in place at Cervia Air Base, Italy. F-15E Strike Eagles had also deployed to Aviano Air Base, Italy and were assigned to the 31st Air Expeditionary Wing.

The F-15C/Ds immediately began to fly combat air patrols, shooting down two Serbian MiG-29s with AIM-120 AMRAAMs on March 24, the first night of the air campaign. Two more MiG-29s were downed on March 26, also with AIM-120s. NATO controlled the skies throughout the remainder of Operation Allied Force. A total of six Serbian aircraft were shot down, four by F-15Cs of the 493rd Fighter Squadron, 48th Fighter Wing, one by an F-16 of the 78th Fighter Squadron, 20th Fighter Wing, and one by a Dutch Air Force F-16. One of the F-15Cs (84-0014) had also scored an aerial victory during Desert Storm, making it the first USAF aircraft in history to score an aerial victory in two different combat operations. The F-15E Strike Eagles, flying from Aviano AB flew ground attack missions using AGM-130 missiles and laser guided bombs. They attacked targets in both Kosovo and Serbia.

A peace agreement between NATO and Serbia was signed on June 9, 1999, following heavy B-52 raids on Yugoslav army units in Kosovo. Once it became clear that the Serbs had complied with the terms of the settlement, and had withdrawn from Kosovo, the massed air armada in Europe began a quick process of redeployment. The 26 deployed 48th Fighter Wing F-15Es returned to RAF Lakenheath from Aviano Air Base, Italy on June 23 and 24, 1999. The 18 deployed F-15C/Ds from the 48th Fighter Wing returned to RAF Lakenheath form Cervia Air Base, Italy on June 25 and 26, 1999.

OPERATION ALLIED FORCE UNITS

31st AIR EXPEDITIONARY WING

When Operation Allied Force Began on March 24, 1999, the 31st Air Expeditionary Wing (AEW) was in place at Aviano Air Base, Italy. The 31st AEW controlled a majority of the U.S. fighter aircraft deployed for Operation Allied Force. The F-15E Strike Eagle Squadron assigned to the 31st AEW was the 494th Expeditionary Fighter Squadron which had deployed to Aviano AB from their home base of RAF Lakenheath. The 494th FS deployed 20 F-15E aircraft and were augmented with an additional six F-15Es from their sister squadron the 492nd, bringing the total of F-15Es deployed to Aviano Air Base to 26. All 26 aircraft returned to RAF Lakenheath on June 23 and 24, 1999.

48th AIR EXPEDITIONARY WING

During Operation Allied force the 48th Air Expeditionary Wing (AEW) remained in place at its home base of RAF Lakenheath, United Kingdom. When the 493rd FS and 494th FS deployed to their forward locations, the 492nd FS, one of the 48th FW's two F-15E Strike Eagle Squadrons, remained at RAF Lakenheath. Augmenting Operation Allied Force, the 492nd Expeditionary Fighter Squadron flew their first combat missions from RAF Lakenheath on May 3, 1999.

501st EXPEDITIONARY OPERATIONS GROUP

The 501st Expeditionary Operations Group (EOG) was in place at Cervia Air Base, Italy. The 501st EOG controlled Cervia Air Base's allied operations during Operation Allied Force. The 493rd Expeditionary Fighter Squadron was F-15C/D Eagle Squadron assigned to the 501st EOG. It had deployed to Cervia AB from its home base at RAF Lakenheath. A total of 18 F-15C/Ds were deployed to Cervia Air Base from RAF Lakenheath. All 18 aircraft returned to RAF Lakenheath on June 25 and 26, 1999.

OPERATION ALLIED FORCE F-15 AERIAL VICTORIES					
DATE	TYPE	SERIAL #/TC	TYPE DEST.	WEAPON	WING/SQDN
03-24-99	F-15C	86-0159/LN	MiG-29	AIM-120	48FW/493FS (501EOG/493EFS)
03-24-99	F-15C	86-0169/LN	MiG-29	AIM-120	48FW/493FS (501 EOG/493EFS)
03-26-99	F-15C	84-0014/LN	MiG-29	AIM-120	48FW/493FS (501 EOG/493EFS)
03-26-99	F-15C	86-0156/LN	MiG-29	AIM-120	48FW/493FS (501 EOG/493EFS)

Right: 84-0014 MARCH 26, 1999 MiG-29 LN Tail Code
By shooting down a Serbian MiG-29 on March-26, 1999, 84-0014 became the first USAF aircraft to have scored an air to air kill in two different conflicts. 84-0014 had also shot down an Iraqi SU-22 on March 20, 1991 while assigned to the 53rd TFS, 36th TFW. In keeping with NATO policy, the names of the pilots shooting down the MiG-29s have not been released. (Neil Dunridge)

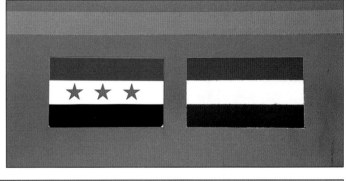

Below: (Neil Dunridge)

86-0156 MARCH 26, 1999 MiG-29 LN Tail Code (Brian C. Rogers)

86-0159 MARCH 24, 1999 MiG-29 LN Tail Code (Neil Dunridge)

86-0169 MARCH 24, 1999 MiG-29 LN Tail Code (Joe Sadler)

CHAPTER EIGHT

Aircraft Description

The F-15 is a high-performance, all-weather fighter capable of attaining airspeeds in excess of Mach 2.5. The air superiority versions' (F-15A/B and C/D) primary mission is aerial combat with the additional task of performing ground attack missions. The F-15E is a true dual role fighter having the air-to air capabilities of the F-15C/D plus air to ground capabilities similar to the F-111F. Radar guided and Infra-Red seeking air-to-air missiles and an internally mounted 20mm gun are the F-15s air-to-air weapons. The F-15s are capable of carrying all types of US gravity weapons, with the F-15E also capable of carrying and guiding most of the USAF guided air-to-ground weapons. The aircraft is powered by two Pratt & Whitney F100 turbofan engines with afterburners providing a high thrust-to weight ratio. Aircraft appearance is characterized by a high mounted, swept-back wing; twin vertical stabilizers; and large left and right horizontal stabilators. The light, high-strength structure is designed to a severe fatigue spectrum for ruggedness and long life. The major aircraft systems are designed and located for survivability, maintainability and reliability. A jet fuel starter provides self-starting of the engines and enhances the concept of a self-sustaining aircraft.

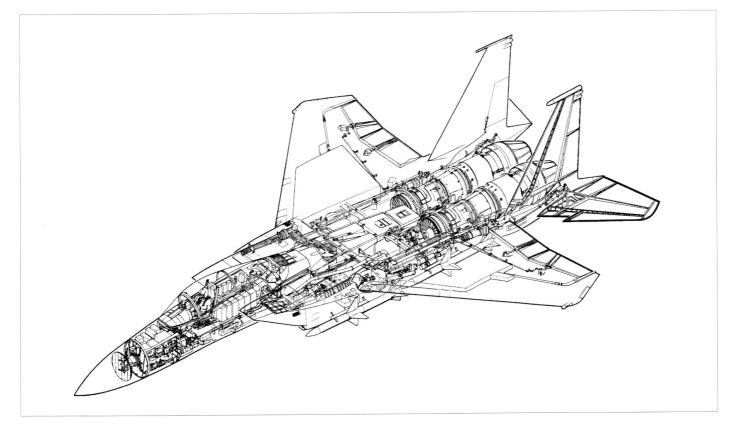

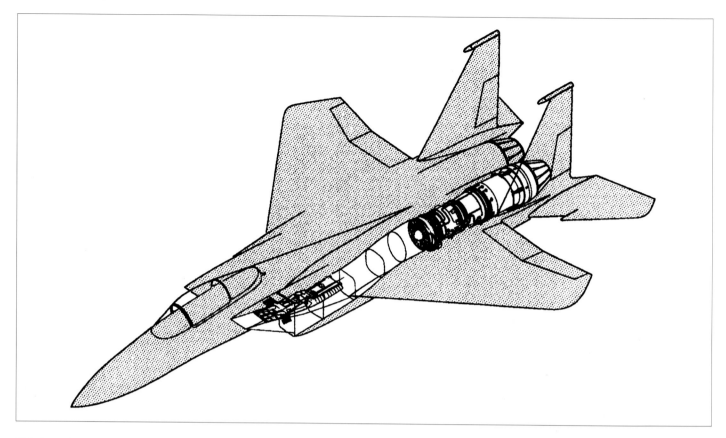

ENGINE AND INLETS

INLETS

There are two independent but identical inlet systems, one for each engine. Each system consists of three variable ramps, a variable diffuser ramp and a variable bypass door. The entire inlet rotates about a pivot on the lower cowl lip to provide optimum airflow at all angles of attack.

Variable Ramp
The variable ramps provide air at optimum subsonic flow to the face of the engine throughout a wide range of aircraft speeds. Ramp position is controlled by the air inlet controller.

Bypass Door
The bypass door controls the inlet duct Mach number by opening automatically to bypass excess air.

Air Inlet Controller
An inlet controller, one for each inlet, utilizes angle of attack, total temperature, ramp actuator position feedback, aircraft Mach number and inlet air Mach number to automatically schedule the ramps and bypass door throughout the aircraft envelope.

ENGINE

The F-15s are powered by two F100 turbofan engines with afterburners. The F100, in the 25,000 pound thrust class, utilizes a three-stage fan with variable inlet guide vanes designed to provide optimum airflow and maximum operational stability throughout the flight envelope. This engine has a 10-stage compressor with three stages of variable stators for optimum airflow scheduling. The F100 is designed to be smokeless. This has been achieved by an annular ram induction burner which concentrates combustion at the burner front end. Extremely high operating temperatures are achieved through the use of an air cooled, two-stage high pressure turbine. The multi-zone afterburner fuel system provides soft lightoff and smooth transient thrust increases from minimum to maximum afterburner levels throughout the flight envelope. The engine exhaust nozzle is a lightweight fully variable convergent/divergent nozzle. The fully-dilating nozzles control the mass flow of air from the engine exhausts. The dilating nozzles were initially fitted with covers called turkey-feathers. These were found to be unnecessary, and because cracking of the turkey feathers became common, they were later removed. Engine maintenance procedures are faster, lower in cost, and can be accomplished at lower maintenance levels because the engine is designed for modular maintenance. This concept allows replacement of major engine assemblies, rather than the customary tear-down and replacement of many separate parts.

All F-15A/Bs and most F-15Cs were delivered with Pratt & Whitney F100-PW-100 turbofans. The F-15C/Ds were later re-

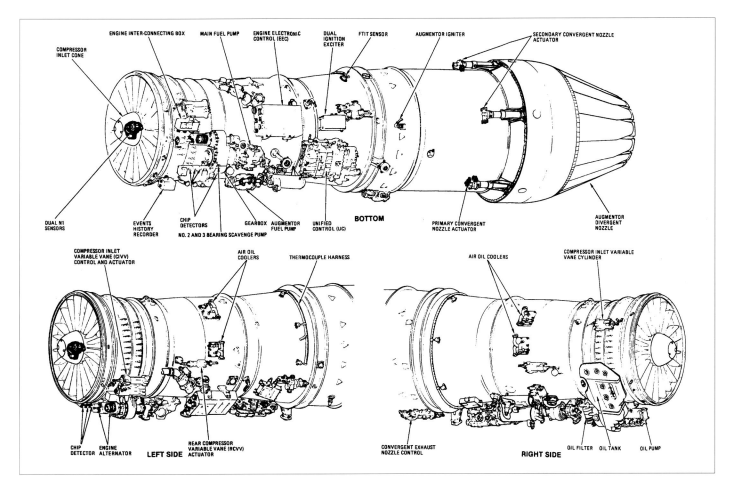

engined with more reliable but slightly lower rated F100-PW-220 engines (maximum afterburning thrust reduced from 23,830 to 23,450 pounds). This engine was first tested on F-15A 71-0287. The new engine was installed starting with F-15Cs on the production line in November 1985, and operational introduction took place in the spring of 1986. The PW-200 engines allowed F-15 pilots to rapidly move both throttles from Mil to Max AB and get full thrust from the engines four seconds later without the worry of engine stalls or RPM hang ups.

AIRCRAFT FUEL SYSTEM

Fuel is carried internally in four interconnected fuselage tanks, and two internal wing tanks. External fuel is carried in three 600 gallon external jettisonable tanks. All tanks may be refueled while the aircraft is on the ground through a single pressure refueling point, or while airborne, through the air refueling receptacle on the left wing shoulder.

The F-15C/D and F-15E had updates to the fuel system. Internally, the F-15C/D and E have additional wing leading and trailing edge tanks, and additional tanks in the central fuselage not included in the F-15A design. They also have added the capability of carrying FAST (Fuel And Sensor Tactical) packs attached to the side of the

fuselage under the wings. The tanks conform to the aerodynamic shape of the side of the fuselage, and as a result had very little adverse aerodynamic effect and very little degradation in performance. The FAST packs are now referred to as Conformal Fuel Tanks (CFTs). Each CFT carries an additional 849 US gallons of fuel. The CFTs were first tested on the second F-15B (71-0291) on July 27, 1974. With the CFTs installed, the F-15 does not loose any weapon carrying capability. The CFTs have weapons stations mounted in the same relative position as on the bare fuselage. The CFT weapons stations have the capability of carrying AIM-7 Sparrow or AIM-120 AMRAAM missiles, bombs, or air-to-surface missiles weighing up to a total of 4400 pounds. F-15Es are seldom seen without their CTFs installed. Although the CFTs alone carry slightly less fuel than the normal three external fuel tanks, they permit the aircraft to be flown at considerably higher speeds.

The fuel transfer system is completely automatic and provides: automatic backup gravity transfer if normal transfer fails, full feed tanks for all engine power settings, and automatic external fuel transfer sequencing. Engine feed tanks are self-sealing for protection from up to 50 caliber projectiles. All internal tanks have reticulated foam for fire/explosion protection. Whenever possible, the fuel lines have been routed through the fuel tanks. Those fuel lines routed outside of the tanks have been covered with a self-sealing material.

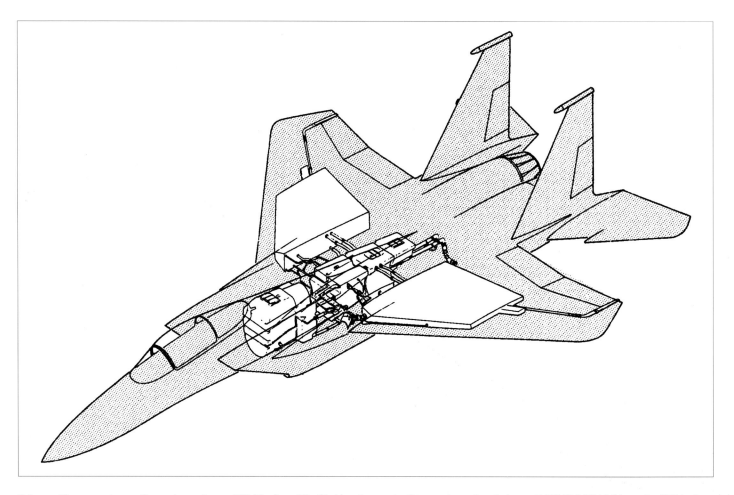

Below and four opposite top: These photos show a CFT (Conformal Fuel Tank) on its carrying fixture prior to installation on F-15E 87-0176. Visible on the CTF in these photos are the six stores ejector racks. (All Don Logan)

This photo shows a CFT installed on the right fuselage side. (Don Logan)

AIRFRAME MOUNTED ACCESSORY DRIVE

The F-15's generators and hydraulic pumps are attached to the Airframe Mounted Accessory Drive unit (AMAD). This unit receives power from the engines via two drive shafts. The AMAD contains four hydraulic pumps and two Integrated Drive Generators (IDG). The hydraulic pump output is 3,000 PSI. The IDCs provide AC power and are rated at 40/50 KVA each.

The Jet Fuel Starter (JFS), a self-contained auxiliary power unit, is mounted between the engines and is connected to the AMAD for ground power. The JFS gives the F-15 a self-starting capability. Starting power for the JFS is provided by a hydraulic motor that is driven by either of two hydraulic accumulators. The JFS also provides electrical and hydraulic power for ground servicing, maintenance and weapons loading.

ELECTRICAL POWER SUPPLY SYSTEM

The electrical power supply system consists of two AC generators, two Transformer-Rectifiers (T-R), an emergency AC/DC generator, and a power distribution system. External electrical power can be applied to the system on the ground. The JFS provides electrical power to part of the electrical system during engine start without external power.

AC Electrical Power

The two AC generators are each mounted on a Constant Speed Drive (CSD) which converts the variable input speed supplied by the engines to the constant RPM required by the generators. A CSD and generator with their integrated oil cooling and lubricating system is called an Integrated Drive Generator (IDG). The IDG oil is independent of the engine oil system. If one of the generators fails, the other generator picks up the entire load. A protection system within the generator control unit guards against damage due to electrical discontinuities.

DC Electrical Power

Two transformer-rectifiers convert AC power to DC power. The output of the T-Rs is connected in such a manner as to prevent a short in the bus of one T-R from damaging the other. If one T-R fails, the other one will pick up the entire load.

Emergency Generator

A hydraulic-motor-driven emergency AC/DC generator is provided for a back-up source of electrical power. The emergency generator is separate from the primary electrical system and is capable of operating equipment required for return and safe landing of the aircraft.

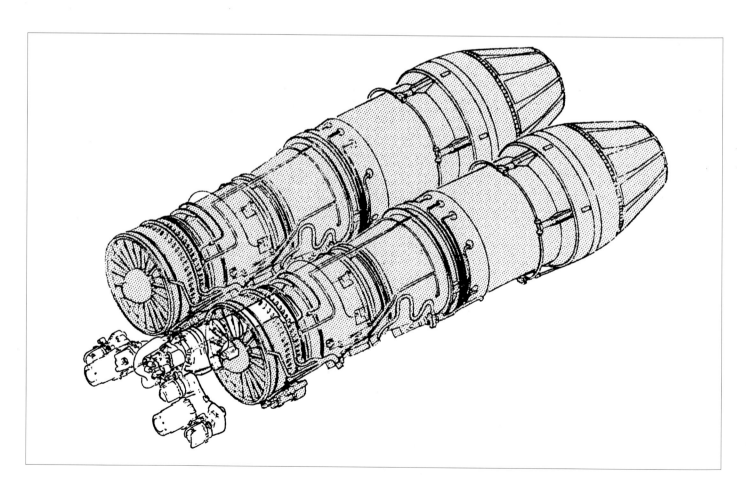

HYDRAULIC POWER SUPPLY SYSTEM

Hydraulic power is supplied by four hydraulic pumps to three separate 3,000 P.S.I. systems. These systems are: power control No. 1, power control No. 2 and utility. Reservoir Level Sensing (RLS) and Return Pressure Sensing (RPS) are two innovations incorporated in the hydraulic system that enhance survivability.

Return Pressure Sensing (RPS)
RPS detects a pressure drop in a subsystem, such as an aileron, stabilator, or rudder, and prevents further flow to that subsystem. If a leak develops in an aileron actuator, the RPS senses a pressure drop and shuts off the flow of hydraulic fluid to that actuator so that the entire hydraulic system is not lost.

Reservoir Level Sensing (RLS)
Each of the hydraulic systems has been divided into two or more separate circuits. RLS is used to detect a leak in one of the circuits and shut that circuit off so that the entire hydraulic system will not be lost. Because of RLS, a leak in one of the hydraulic systems causes only a part of that system to be lost.

PRIMARY FLIGHT CONTROL SYSTEM

The aircraft primary flight control system consists of conventional ailerons, twin rudders, and left and right symmetrical/differential stabilators. The control surfaces are positioned by hydraulic actuators that receive mechanical and/or Control Augmentation System (CAS) inputs. The function of the CAS is to measure control forces exerted by the pilot, measure aircraft response to these commands and then displace the control surfaces to the position that makes the airplane respond in the desired manner. The CAS has a fail-safe feature that prevents large unwanted control inputs. Ailerons are powered by two hydraulic systems and all tail surfaces are powered by three hydraulic systems. Artificial feel systems provide simulated aerodynamic forces to the control stick and rudder pedals. The feel systems have actuators which move the entire control surfaces.

Roll Control
Ailerons and differential deflection of the stabilators are used to roll the aircraft. Safety spring cartridges allow operation of a single aileron-stabilator combination if the other side should become jammed. If both ailerons become jammed, the differential stabilators provide lateral control for moderate flight maneuvers, including landing. If all mechanical linkages are severed, the aircraft can be

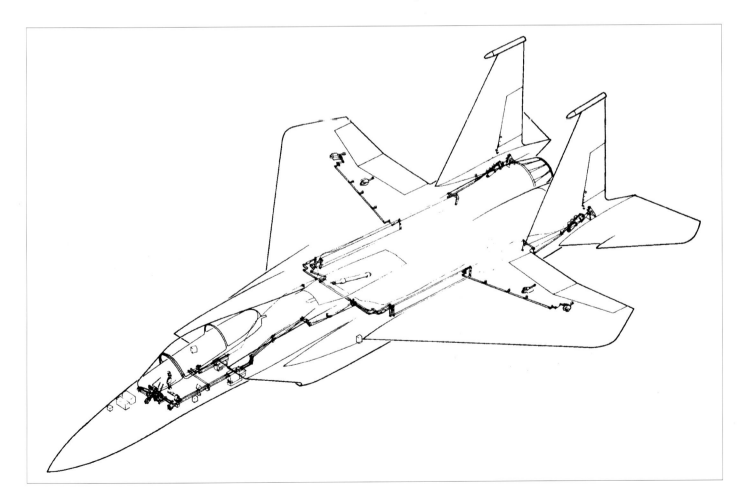

recovered using differential stabilators controlled by the electronic CAS.

Pitch Control

Pitch control is provided by the stabilators. If the mechanical linkage becomes jammed, the stabilators through the electronic CAS, will provide longitudinal control for moderate flight maneuvers, including landing.

Yaw Control

Two rudders are used for yaw control. Rudders are actuated by the rudder pedals and the flight control stick through the aileron-rudder interconnect. If the rudder linkage jams, safety spring cartridges allow the rudders to be moved through the CAS.

LANDING GEAR

The landing gear system consists of two main landing gear and a nose landing gear. Each landing gear has one wheel. The main landing gear retract forward into the fuselage, turning 90 degrees as they retract to lie flat in wheel wells underneath the fuselage. The landing gear track is narrow (only nine feet) because a wider track landing gear would have incurred an unacceptable increase in aircraft weight. The narrow track causes some problems during crosswind landings. The upwind wing tends to rise, causing the aircraft to weather vane (yaw) into the wind and then drift downwind. The nose landing gear retracts forward into a wheel well underneath the pilot's cockpit. The nose wheel is steerable through 15 degrees left and right. A retractable arresting hook is located on the bottom of the aircraft between the engine exhausts. The hook is designed to stop the aircraft in emergency situations by snagging an arresting cable stretched across the runway.

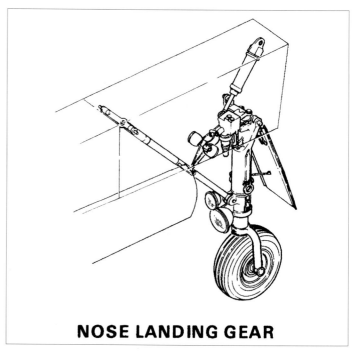

NOSE LANDING GEAR

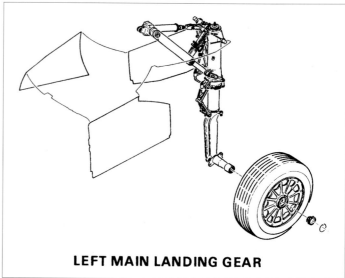

LEFT MAIN LANDING GEAR

Left: All F-15s have an extendible arresting hook used to stop the aircraft during landing or takeoff emergencies. (Don Logan)

COCKPIT FACILITIES

The pilot's cockpit is mounted high on the forward fuselage behind a one-piece windshield. The canopy itself is a single transparency with only one transverse frame. It is hinged at the rear and opens in a clamshell-type fashion. The cockpit canopy offers excellent 360 degree visibility. The cockpit length of the single seat and two seat versions is the same, with avionics occupying part of the space aft of the pilot's ejection seat in the single seat versions. The canopy shape differs with the two seat versions allowing for rear seat and occupant clearance

The aircraft is provided with a McDonnell Douglas ACES II ejection seat, with zero-zero capability. At zero airspeed, the catapult fires within 0.3 seconds, followed by the rocket sustainer in 0.45 seconds, separation of the pilot from the seat after 1.3 seconds, and opening of the parachute pack in 2.3 seconds.

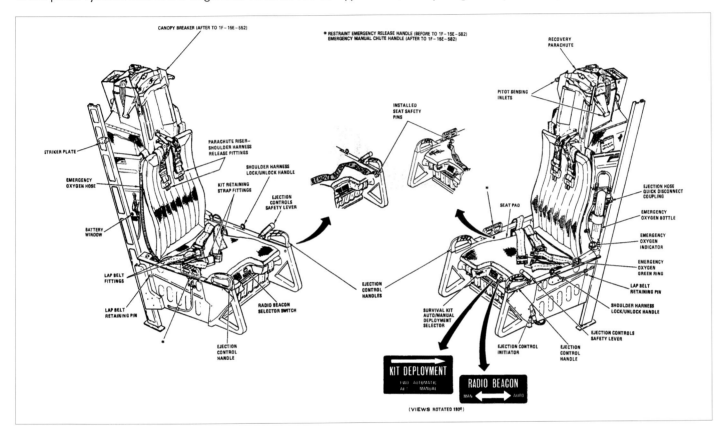

F-15 A/B & C/D AIR TO AIR WEAPON SYSTEM

The air-to-air weapon system consists of an AN/APG-63 radar in F-15A/Bs and early F-15C/Ds. The later F-15C/Ds received the AN/APG-70, with some aircraft modified with the new upgraded AN/APG-63(V)1. All the F-15 radars are pulse doppler radar. In addition to the radar the weapon system consists of a central computer, an armament control panel, a radar display and a heads-up display (HUD). Built by Hughes, the APG-63/APG-70 radar is known for its flexibility. With a maximum detection range of 100 miles, it can look up and detect high-speed, high-altitude targets or look down and detect low altitude targets. This radar has several different air-to-air modes, long-range search, velocity search mode, and short-range search modes. There is also a non pulse-Doppler mode which is useful only when looking up. A beacon mode interrogates other aircraft IFF transponders. A manual tracking mode can be used as a backup if the radar is not automatically tracking well, and a "sniff" mode which detects jamming and emits tiny bursts to minimize self-illumination.

The central computer is used to relieve the pilot's work load. It continuously performs built-in tests of the avionics system to alert the pilot if his system becomes degraded. It performs many of the radar switching functions automatically. It computes the information necessary to fire the missiles and the gun. The armament control panel is used to monitor the status and arm the gun and missiles. The radar display provides all attack symbology necessary to acquire, track, lockon, and fire missiles. The HUD provides symbology to navigate, fly instruments, and acquire, lockon and track targets.

In order to operate the weapon system with a minimum of effort, key radar and armament control switches are located on the stick and the throttle. Target acquisition, target lock-on, missile

Visible in this photo are four AIM-7 Sparrow Missiles on the fuselage and four AIM-9 Sidewinder missiles on the wing pylons. (USAF via Marty Isham)

selection and missile firing are all accomplished using these switches. Flight parameter and target information is displayed on the head up display (HUD). Using the HUD the pilot is able to observe the real world and the necessary attack information without having to look inside the cockpit.

As compared to the F-15A, significant improvements were made to the avionics suite of the F-15C. The APG-63 radar of the F-15C is equipped with a Programmable Signal Processor (PSP) which is a special-purpose computer that controls the radar modes through software. This permits more rapid switching of the radar between modes for maximum operational flexibility.

F-15 A/B & C/D AIR TO GROUND WEAPON DELIVERY SYSTEM

The F-15 was designed as an air superiority fighter, although very rarely functioning in a ground-attack role, the F-15A/B/C/D Eagles have a secondary air-to-ground capability. Up to 16,000 pounds of bombs, fuel tanks, and missiles can be carried. The underwing py-

lons can each accommodate a multiple ejector rack which can carry six 500-pound bombs. The bomb racks can be installed on the underwing pylons without disrupting the normal carriage of Sidewinder missiles. Air-to-ground stores can also be carried on the underfuselage centerline.

The F-15A/B/C/D can carry and deliver laser-guided bombs such as the GBU-10E/B Paveway II or the GBU-12D/B Paveway II. However, they do not have the capability of guiding these weapons, and must rely on laser designators carried by other aircraft or by personnel on the ground.

The air-to-ground weapon delivery system is optimized for one man operability. The work load has been minimized by automating weapon delivery tasks. As with the air-to-air modes, pilot effectiveness for air-to-ground modes has been increased by the use of the HUD and by mounting all essential weapons controls on the stick and the throttle.

The air-to-ground avionics system consists of the AN/APG-63/70 radar, a central computer, and an inertial navigation system (INS). The air-to ground modes use the same cockpit controls and displays as do the air-to-air modes. The radar provides slant range information for the two computed delivery modes and the direct mode. The central computer helps to minimize the pilot's work load by performing switching functions, computing steering and release information, and controlling the radar. The armament control panel is used to select, precondition, arm, release and jettison weapons. The radar display provides head-down attack symbology and sensor video. This includes radar video (ground map and air-to-ground ranging) and Electro-Optical guided weapon video. The head up display provides attack symbology and steering information. All information necessary during the combat profile is projected onto a combining glass in front of the pilot, so that he can keep his eyes out of the cockpit during a delivery pass. The INS is the primary source of aircraft pitch, roll, heading, acceleration, velocity and present position information. This data is used to compute steering and weapon delivery information. The INS can store up to 12 different destinations and offset points.

MULTI-STAGE IMPROVEMENT PROGRAM (MSIP)

The Multi-Stage Improvement Program (MSIP) was a joint program carried out by McDonnell Douglas (now Boeing) and the USAF's Warner Robins Logistics Center in Georgia. Almost all F-15A, B, C, and D versions went through the program. However,

Stores	8	7/7C	RC	6/6C	5	4/4C	LC	3/3C	2	MSIP Load (5-Sta)	MSIP Load (3-Sta)	Exist Load (3-Sta)
Air-to-Air Missiles												
– AIM-7F		1/1		1/1		1/1		1/1		4	4	4
– AIM-9J, AIM-9P, AIM-9L	2								2	4	4	4
– AIM-120A Provisions	2	1/1		1/1		1/1		1/1	2	8	8	0
– AIM-7M		1/1		1/1		1/1		1/1		4	4	0
– AIM-9M	2								2	4	4	0
General Purpose												
– MK-82 LDGP/R	6				6				6		18	18
	4		4		6		4		4	22		
– MK-84 LDGP	1		1		1		1		1	5	3	3
Guided Bombs												
– GBU-8/B	1		1		1		1		1	5	3	3
– GBU-10A/B, C/B	1		1		1		1		1	5	3	3
Dispensers												
– CBU-52B/B, 58/B, 71/B	4		3		4		3		4	18	12	12
– MK-20 Rockeye	4		4		6		4		4	22		
	6				6				6		18	18
Training												
– SUU-20B/A	1		1		1		1		1	5	3	3

- 5-Station Capability
  - PACS Hardware
  - PACS OFP
  - CC OFP
  - Aircraft Wiring to CFT
- CFT Baseline Per ECP 1377R1
- CFT A/G Adapters Not Included in MSIP

some of the very early As (from FY 1973, 1974, and 1975) were not upgraded under MSIP and were retired or made available as gate guards or donated to museums.

Under MSIP, upgrades were progressively incorporated onto the production line and then retrofitted to earlier production F-15Cs. One of these upgrades involved an improvement of the capabilities of the AN/APG-63 radar and fire control system. The memory capability of the APG-63 radar fire control system was increased from 96K to 1000K and the processing speed was trebled. A Programmable Armament Control Set (PACS) was installed. Also, MSIP aircraft were fitted with the wiring needed to give them the capability of carrying and launching the AIM-120 AMRAAM missile, introduced into service in the early 1990s.

Another part of the MSIP is the Seek Talk program, which was designed to reduce the vulnerability of the F-15's UHF radios to enemy jamming by introducing spread spectrum techniques and the use of a null steering antenna. Yet another was the Joint Tactical Information Distribution System (JTIDS), which provides a high-capacity, reliable, and jam-proof information distribution system between various elements of deployed forces and command and control centers. Another aspect of the MSIP is the integration of the F-15 with the Global Positioning Satellite (GPS).

MSIP replaced the analog computers of the F-15A/B with digital computers and upgrades the digital computers of the F-15C/D. MSIP uses a weapons panel and a cathode ray terminal similar to that found on the F-15E. The F-15C/D were fitted with chaff/flare dispensers behind the nosewheel door. The A models that went through the MSIP were not be fitted with the conformal fuel tanks of the C, but are otherwise indistinguishable from the C models.

F-15E WEAPON DELIVERY SYSTEM

The heart of the F-15E's Weapon Delivery System is the AN/APG-70 Synthetic Aperture Radar (SAR). SAR imagery sharpens the details of the radar ground map returns and provides an overhead map type view of the ground returns. In addition to the full air-to-air capability of the F-15C, the radar provides a display of ground targets that are of higher resolution and taken from further away than the images produced by non SAR radars. Roads, bridges, and airfields can be identified as far away as 100 miles, and as the F-15E nears the target, image resolution becomes progressively sharper, and smaller targets such as trucks, aircraft, and tanks can be distinguished.

The LANTIRN (which is an acronym standing for Low-Altitude Navigation and Targeting, Infra-Red for Night) system consists of two pods, one carried underneath each air intake. The right side pod is used for navigation and contains a Forward Imaging Navigation Set (FINS) which can be used to display a high-quality video image of the oncoming terrain on the pilot's heads-up display, enabling high-speed low-level flights to be made at night under clear weather conditions. The navigation pod also carries a terrain-following radar which is also effective in bad weather. The pilot can manually respond to cues from the system or can couple the sys-

tem to the flight controls for "hands-off" automatic terrain-following flight at altitudes as low as 200 feet above the ground. The left side pod is a targeting pod which contains a high-resolution tracking FLIR (Forward-Looking, Infra-Red), a missile boresight correlator, and a laser designator. The boresight correlator is used to guide the Maverick air-to-surface missile and the laser designator is used for weapons such as laser guided bombs that home in on reflected laser light.

The F-15E uses conformal fuel tanks like the F-15C/D. The air-to-ground weapons load on the F-15E version of the CFT was raised to a maximum of 23,500 pounds by adding six stub pylons on the corner of each conformal fuel tank for external ordnance. The F-15E retains the 20-mm M61A1 cannon of the F-15D, although the ammunition capacity is reduced to 450 rounds. The F-15E retains the full air-to-air capability of the F-15C/D version and can carry AIM-120 AMRAAM or AIM-7M Sparrow medium-range missiles and AIM-9M Sidewinder infrared missiles.

F-15E LANTIRN SYSTEM

The LANTIRN is an imaging infrared system which provides the F-15E with a day/night under the-weather attack capability to deliver a wide variety of conventional and precision-guided munitions at night using day-like tactics. LANTIRN consists of two pods, the AN/AAQ-13 Navigation Pod and the AN/AAQ-14 Targeting Pod.

AN/AAQ-13 NAVIGATION POD

The Navigation Pod is fully integrated with the aircraft's avionics computer to provide navigation and target area acquisition information. It can also be used to deliver unguided, free-fall ordnance with the aircraft's onboard fire control computer. The Navigation Pod has two main components, the Forward Imaging Navigation Set (FINS) sensor and the Ku-band Terrain Following Radar (TFR). The FINS, a Wide Field of View (WFOV) Forward Looking Inra-Red (FLIR) system, provides the crew with an I-R image of the terrain and airspace in front of the aircraft. The FINS includes a look-into-turn mode which enables the crew to look ahead of the turn while turning, and a snap-look mode which provides enhanced left, right, up, and down viewing control while flying straight and

The AN/AAQ-13 Navigation Pod is carried under the right intake on F-15E aircraft. (Don Logan)

level. The FINS functions include video polarity control (changes the display between normal black and white video, and reversed video similar to a photo negative), video gain and level adjustment options, and a gray scale capability for manual gain and level setting. The TFR features includes terrain following, obstacle warning, and limited IMC (Instrument Meteorological Conditions) flying. The Selected Clearance Plane settings between 200 and 1,000 feet AGL are available in the normal mode. In addition to normal operation of the TFR, other modes are available for specific operating conditions. These modes are weather (WX), low probability for intercept (LPI), electronic counter-countermeasures (ECCM), and very low clearance (VLC). Within the Navigation Pod the TFR and FINS are functionally independent of each. However, operationally, the crew depends on a cross-check of the information from both systems to perform night, low-level navigation tasks.

AN/AAQ-14 TARGETING POD

The targeting pod can acquire targets at sufficient slant range to cue, lockon and fire IR Mavericks, and designate stationary targets for the delivery of laser-guided bombs. The Targeting Pod also enhances unguided conventional weapons delivery capabilities. Targeting Pod functions included FLIR imaging, laser designation and ranging, accurate pointing and target tracking, missile boresight correlation, and IR Maverick hand off for subsequent manual launch. The optical system is inertially stabilized by positioning the IR optics to either scan a target area or to track a recognized target. The IR optics are mounted on pitch and yaw gimbals in forward portion of the pod nose. The Target Acquisition FLIR (TAF) is mounted in the aft portion of the pod nose section. It corrects for IR scene rotation caused by the motion in the stabilization system. This allows the IR image displayed to the crew to always appear right side up. The whole nose section rotated or "rolled" to maintain track on a target while the TAF assembly was "derolled' to preserve proper orientation of the IR image.

F-15E AN/AXQ-14 DATA LINK POD

When controlling GBU-15 glide bombs or AGM-130 missiles after launch, the controlling aircraft carries a AN/AXQ-14 data link pod. The F-15E carries the pod on the centerline station. Command

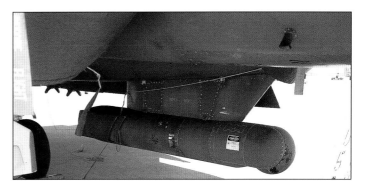

Below: The AN/AAQ-14 Targeting Navigation Pod is carried under the left intake on F-15E aircraft. (Don Logan)

and control signals from the Weapon Systems Officer (WSO) are transmitted from the aircraft to the weapon, and data and target video are received by the pod from the weapon and displayed to the WSO.

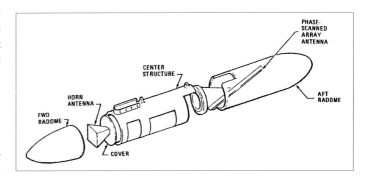

AN/AXQ-14 Data Link Pod

337

CHAPTER NINE

F-15 Weapons

F-15 AIR-TO-AIR WEAPONS

There are four weapons making up the F-15s Air-to-Air armament, the AIM-9 Sidewinder, the AIM-7 Sparrow and later the AIM-120 AMRAAM missiles, and the internal 20mm M-61 Vulcan cannon for short-range encounters.

AIR LAUNCHED MISSILES

GUIDANCE

All types of air-to-air missiles carried by the F-15 are guided missiles. Guided missiles may either home to the target, or follow on a non-homing course. Non-homing guided missiles are either inertially guided or reprogrammed. Homing missiles may be active, semiactive, or passive.

ACTIVE

An active missile carries the radiation source on board the missile. Radiation from the missile is emitted, strikes the target, and is reflected back to the missile. The missile then is self-guided on this reflected radiation. The AIM-120 AMRAAM is an active homing missile.

PASSIVE

A passive missile uses radiation originated by the target or by some source not a part of the overall weapon system. The AIM-9 Sidewinder is a passive homing missile. Typically, for air-to air, this radiation is in the infrared (IR) region.

SEMI-ACTIVE

A semiactive missile has a combination of active and passive characteristics. A source of radiation is part of the system but is not carried in the missile. The source (usually the radar system of the launching aircraft) radiates energy to the target from which the energy is reflected back to the missile. The missile senses the ra-

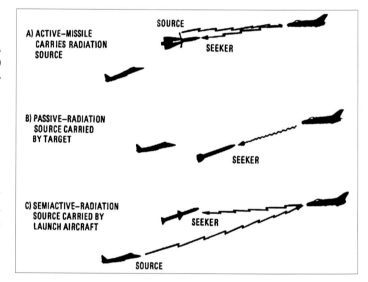

diation and homes to it. The AIM-7 Sparrow is a semi-active homing missile.

AIM-120 AMRAAM MISSILE

The AIM-120A is an all-weather, beyond-visual-range (BVR), radar-guided, air-to-air missile with launch and leave capability. The missile permits launch and maneuver by the launching aircraft at ranges in excess of the AIM-7. It has replaced the AIM-7 as the F-15 primary BVR missile. The AMRAAM has a rail and/or ejection launch capability. The missile guidance system incorporates four guidance modes: (1) an active radar with home-on jam during any phase of flight; (2) command update at long range plus active terminal; (3) inertial plus active terminal if command update is not available; and (4) active terminal with no reliance on aircraft fire control systems at ranges within seeker acquisition range. The AMRAAM is propelled by a solid fuel, reduced smoke rocket motor. The AIM-120B, visibly the same missile as the AIM-120A with AMRAAM Production Enhancement Program (PREP) Block I/II configurations and

338

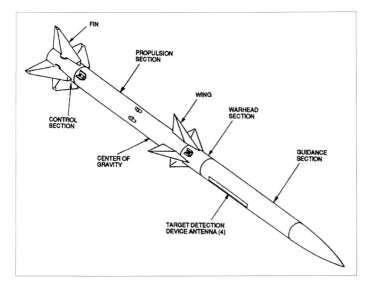

The AIM-120 AMRAAM seen here on the outboard missile rail of an F-15E wing pylon is the Eagle's primary BRV (Beyond Visual Range) missile. The AIM-120 is capable of either rail launch from wing pylons or ejector launch from the fuselage stations (Don Logan)

software reprogrammability added. The AIM-120C, designed for carriage in the internal bay of the F-22, has enhanced hardware and software capability over the AIM120B and has visibly different clipped wing and fins.

The AMRAAM is 11.97 feet long, has a wingspan of 20.7 inches, and a diameter of seven inches. The AMRAAM weighs approximately 350 pounds at launch. It carries a 48-pound high-explosive directed-fragmentation warhead. Maximum speed is about Mach 4, and the maximum range is approximately 35-45 miles.

AIM-9 SIDEWINDER

The AIM-9 Sidewinder missile is a supersonic air-to-air intercept missile. It is a passively-guided missile that guides on infra-red (IR) radiation generated by a target. Because no guidance is required after launch, the pilot may take evasive action immediately after the missile is launched. The Sidewinder is 9.4 feet long, has a tail fin span of 25 inches and a diameter of five inches. The missile has four tail fins on the rear, with a rolleron at the tip of each fin. The rollerons are spun at high speed by the slipstream and, as a result of their high RPM spinning, gyroscopically provide roll stability. The missile is steered by four canard fins mounted in the forward part of the missile just behind the infrared seeker head. The Sidewinder missile has a launch weight of approximately 180 pounds and a maximum effective range of near 10 miles. The blast-fragmentation war-

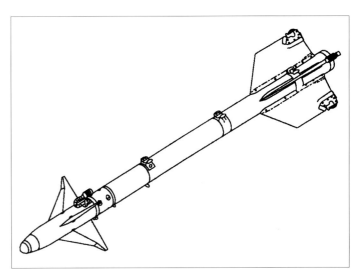

Two below: The AIM-9 Sidewinder seen here on the outboard missile rail of a F-15E wing pylon is a passive infrared missile used for targets at ranges up to 10 miles. The AIM-9 is a rail-launched missile. (Don Logan)

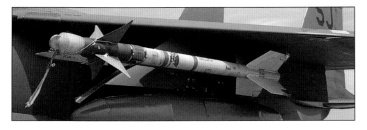

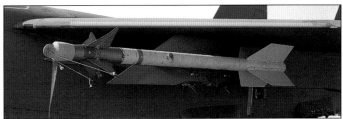

head weighs 21 pounds. The AIM-9 missile consists of four external sections: Guidance and Control Section (GCS), warhead, fuze, and missile body (rocket motor). The AIM-9 missile interfaces with the aircraft through the umbilical cable. The missile has three basic phases of operation; captive flight, launch, and free flight. Power is supplied from the launcher during captive flight. The power is switched to the missile contained thermal batteries during the launch phase and during free flight.

Four AIM-9 Sidewinders (presently AIM-9M) can be carried on the underwing pylons of the F-15, two on each side. Attachment points for Sidewinder missiles rails are on each side of the underwing pylons, enabling a drop tank or a bomb ejector rack to be carried underneath the pylons at the same time as the Sidewinders. Presently the F-15s carry AIM-9Ms. The AIM-9M, introduced in 1982, had an improved capability to distinguish between aircraft and decoy flares. It also has a low-smoke rocket motor so that it is less likely to be seen by the intended target. The AIM-9M missiles are similar in appearance to other AIM-9 versions but are more maneuverable, have an all-aspect capability due to a more sensitive IR detector with a coolant gas system for the IR detector.

AIM-7 SPARROW

The AIM-7 is a supersonic, air-to-air guided missile. The missile can intercept and destroy targets in adverse weather conditions. The AIM-7 is a semi-active missile guided on either continuous wave (CW) or pulse doppler (PD) radio frequency (RF) energy radiated by the launching aircraft and reflected by the target. The missile is guided, controlled, and detonated by the target seeker and flight control sections. The solid propellant rocket motor provides the thrust in a boost-sustain mode. Though being phased out and replaced by the AMRAAM, the AIM-7M is the model used by the F-15. The AIM-7 was responsible for 25 of the 35 F-15 kills scored during Desert Storm. The AIM-7M is 12 feet long and has a launch weight of about 500 pounds. The missile carries a 85-pound high-explosive blast fragmentation warhead. It has two sets of delta-shaped fins; a set of fixed fins at the rear of the missile, and a set of movable fins at the middle of the missile for steering.

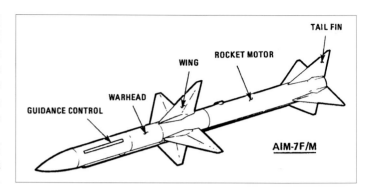

The AIM-7 Sparrows seen here on the missile stations of a CFT were the Eagle's first BRV (Beyond Visual Range) missile. It has since been replaced by the AIM-120 AMRAAM as the Eagle's primary BVR missile. (Don Logan)

The AIM-7 Sparrows are seen here on the fuselage missile stations of an F-15A. (Don Logan)

M61A1 20mm CANNON

For the very closest air-to-air encounters, the F-15 carries a 20-mm M61A1 cannon installed in the right wing leading edge lip, just outboard of the upper air intake. The gun is fed by an ammunition drum containing 940 rounds for the F-15 A/B/C/D and 450 rounds for the F-15E. The drum is located inside the central fuselage just below the hinge of the large dorsal air brake.

The M61A1 20mm Cannon is mounted near the top of the fuselage on the right intake. (Don Logan)

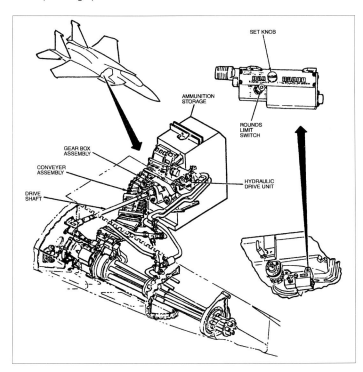

F-15E M61A1 installation

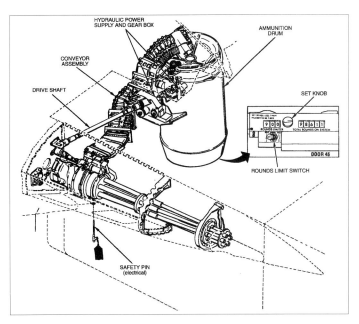

F-15A, B, C, D M61A1 installation

F-15E GRAVITY WEAPONS

The F-15E can presently carry the majority of the gravity non-nuclear weapons in the US inventory including general purpose bombs (MK series and BLU series) and Cluster Bomb Units (CBUs). In addition it can carry and release Laser Guided Bombs, using the LANTIRN for target illumination. The Strike Eagle can also carry, release and guide the longer range GBU-15 Modular Guided Weapon System (MGWS) family of glide bombs and the AGM-130 family of rocket boosted glide bombs.

During summer 1999 Boeing, St. Louis, Missouri was awarded a contract for the development of the Suite 4E+ Operational Flight Program software which allows integration of smart weapons, including the Joint Direct Attack Munition (JDAM GBU-31 series), Joint Standoff Weapon (JSOW AGM-154 series), and Wind Corrected Munitions Dispenser (WCMD CBU-103, CBU-104, and CBU-105) onto F-15Es. The expected completion date for this effort is June, 2002.

This BDU-50 seen on a CFT station is the training weapon used to simulate the B-61 nuclear bomb carried by the F-15E. (Don Logan)

This Tactical Munitions Dispenser is the dispenser used for carrying modern cluster bomblets. When loaded with BLU-97 Combined Effects Munitions (CEMs), it is designated as a CBU-87; when loaded with BLU-91 or BLU-92 GATOR mines, it designated a CBU-89; and when loaded with BLU-108 submunitions, it is designated a CBU-97. (Don Logan)

This photo shows two GBU-12 Paveway II Laser Guided Bombs (LGBs) on the CFT aft weapon stations. The GBU-12 uses a MK 82 500 pound class bomb as the warhead. (Don Logan)

This photo shows a GBU-28 Paveway III Laser Guided Bomb (LGB) on the fuselage center station. The GBU-28 uses a BLU-113 4700 pound class penetrator as the warhead. Early BLU-113s were made from Howitzer gun barrels welded to nose and aft pieces. The later versions of the BLU-113 are now produced from a one-piece forging. (Don Logan)

This photo shows a MK 84 Air Inflatable Retarder (AIR) Hi-Drag 2000-pound class General Purpose bomb. (Don Logan)

This photo shows a GBU-10 Paveway II Laser Guided Bomb (LGB) on the wing pylon. The GBU-10 uses a MK 84 2000 pound class bomb as the warhead. (Don Logan)

This photo shows a GBU-10 Paveway II Laser Guided Bomb (LGB) on the wing pylon and two GBU-12 Paveway II Laser Guided Bombs (LGBs) on the CFT forward weapon stations. (Don Logan)

This photo shows an SUU-20 Practice Bomb Dispenser on the wing pylon below the AIM-9. The SUU-20 can carry up to six practice bombs and four 2.75 inch folding fin rockets. (Don Logan)

GBU-15

The GBU-15 Modular Guided Weapon System (MGWS) is a family of weapons providing accurate guided delivery at an increased range over gravity ballistic bombs. They use either Television or IR guidance systems. The GBU-15 (V)1/B and GBU-15 (V)2/B use the MK84 2000 pound class general purpose bomb, with the GBU-15 (V)31/B and GBU (V)32/B using the BLU-109/B 2000 pound class penetrator bomb. The effective standoff range from the target is increased because it is not necessary to acquire the target before release. The weapon can be remotely controlled before and after release by the Weapon Systems Officer (WSO) through the AN/AXQ-14 Data Link Pod system. The target area and specific aiming point can be located after release using the video image transmitted by the weapon to WSO in the controlling aircraft. The weapon mid course flight path can be adjusted by the WSO, or the weapon can be allowed to continue on its programmed course. Likewise, automatic terminal target tracking with aim point updating can be used, or if desired, the weapon can be manually steered to target impact by WSO commands.

AGM-130

The AGM-130 missiles are a family of air-to-ground missiles with a glide, boost, glide flight profile from release to target impact. Like the GBU-15 family, they use either Television or IR guidance systems. The AGM-130A-1, -2, and -3 use the MK84 2000 pound class general purpose bomb, with the AGM-130C-1, -2, and -3 using the BLU-109/B 2000 pound class penetrator bomb. The effective standoff range from the target is increased by the addition of a rocket propulsion system. The missiles have a capability of both high and low altitude release with both direct and indirect target attack profiles. The indirect mode has a low altitude cruise flight which uses a radar altimeter to maintain clearance above the ground. The target area and specific aim point can be located after release. Some versions also have added a Global Positioning System/Inertial Navigation System (GPS/INS) which gives the missile increased accuracy over a longer range. Like the GBU-15, the AGM-130 can be remotely controlled before and after release by the WSO through the AN/AXQ-14 Data Link Pod system. In the indirect attack mode, the target area and specific aiming point can be located after release using the video image transmitted by the missile to WSO in the controlling aircraft. The missile flight path can be adjusted by the WSO, or the weapon can be allowed to continue on its programmed course. Likewise, the missile's automatic terminal target tracking with aim point updating can be used, or if desired, the missile flight path can be manually adjusted to target impact by WSO commands.

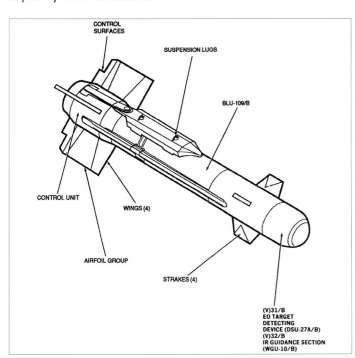

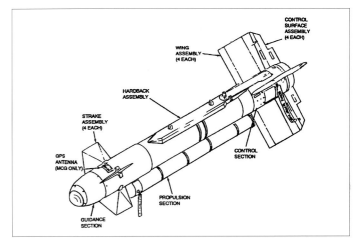

These photos show AGM-130 missiles on F-15E wing pylons. The AGM-130 was fired by F-15s against Iraq during Operation Southern Watch and was also used against Serbian Forces during Operation Allied Force. (Don Logan)

CHAPTER TEN

Paint and Markings

The first 42 F-15s (71-0280/0291, 72-0113/0120, 73-0085/0099, and 73-0108/0114) were painted air-superiority blue. Two aircraft were exceptions; 71-0287 was painted in gloss white for its use as the spin-test aircraft, and 72-0119 (Streak Eagle) which was not painted to reduce weight for time to climb record flights. Air superiority blue consisted of flat AS Blue (FS35450) on the upper surfaces and gloss AS Blue (FS15450) on the lower surfaces. This paint scheme was considered effective in the blue skies over Edwards AFB and Luke AFB, but it was not effective in cloudy skies, such as those expected to be encountered over Europe most of the year.

All the Air Superiority Blue operational aircraft were repainted into Compass Ghost during their first depot-level maintenance period. Initially, tail codes were painted in white like other tactical aircraft of the time. They were later changed to black. Early production aircraft after 73-0100 (or 74-0137 for two-seat aircraft), including foreign deliveries, were painted in Compass Ghost. Project Compass Ghost has been a continuing effort by the military to find the most effective camouflage scheme possible. The standard Compass Ghost F-15 scheme prior to 1990 consisted of two shades of gray, called Light Ghost Gray (FS36320) and Dark Ghost Gray (FS36375), applied in a pattern designed to minimize the reflectance of various contours of the aircraft.

Beginning in 1990, Compass Ghost was replaced by the High and Low Reflectance Gray scheme that uses two new shades of gray (FS36251 and FS 36176). This so-called Mod Eagle scheme has been applied to almost all F-15s as they progress through depot-level maintenance or the MSIP program.

All F-15s have the interior of their air intakes painted in gloss white. The gloss white brightens the interior of the intake and helps to detect foreign objects in the intake. Wheel wells, landing gear struts, and the interior of the speed brake area are also generally painted gloss white. A majority of the interior surfaces are finished with a metallic green coating for corrosion resistance.

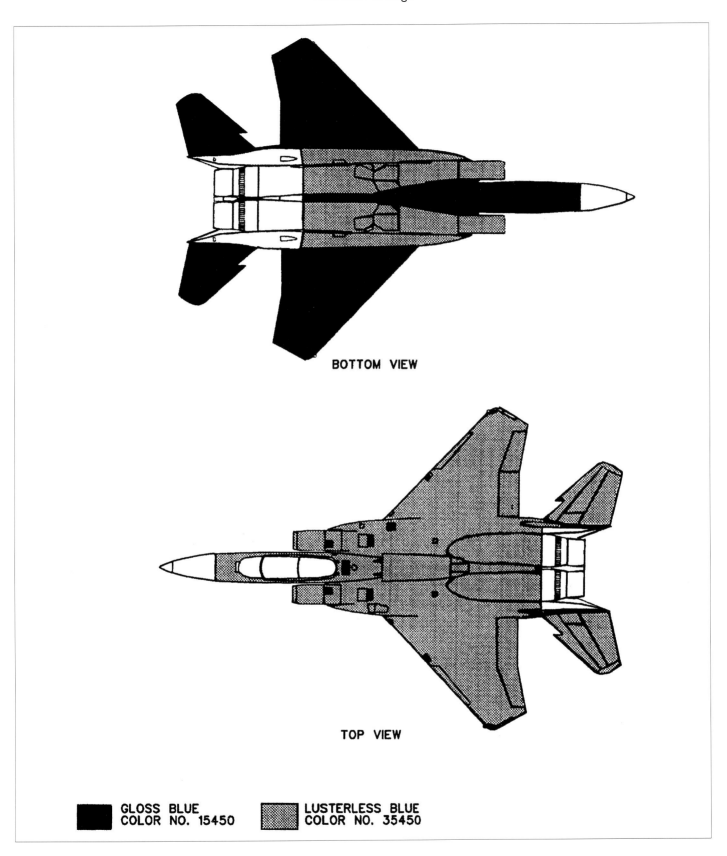

BOTTOM VIEW

TOP VIEW

GLOSS BLUE
COLOR NO. 15450

LUSTERLESS BLUE
COLOR NO. 35450

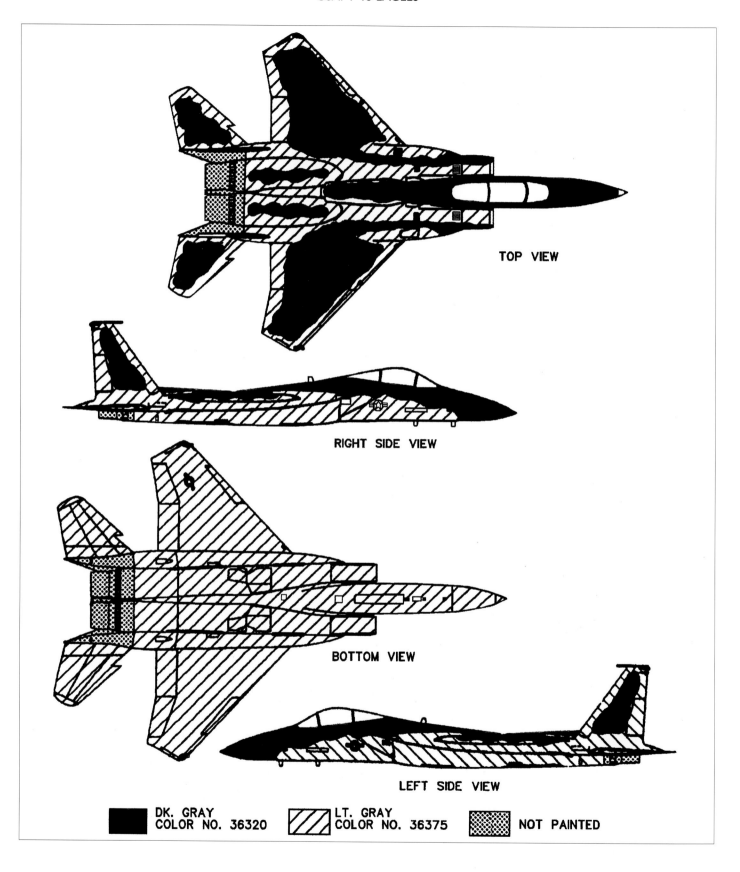

TOP VIEW

RIGHT SIDE VIEW

BOTTOM VIEW

LEFT SIDE VIEW

DK. GRAY
COLOR NO. 36320

LT. GRAY
COLOR NO. 36375

NOT PAINTED

Appendices

Specification of McDonnell Douglas F-15A/B & C/D

Two Pratt & Whitney F100-PW-100 (A/B), F100-PW-200 (C/D) axial-flow turbofans, each rated at 12,420 pounds dry, 14,670 pounds at full military power, and 23,830 pounds with afterburning.

Maximum speed: 1650 mph (Mach 2.5) at 36,000 feet, 915 mph at sea level. Cruising speed 570 mph. Initial climb rate 40,000 feet per minute. Service ceiling 65,000 feet. Maximum unrefuelled range 3450 miles.

Dimensions: Wingspan 42 feet 9 1/2 inches
Length: 63 feet 9 inches
Height: 18 feet 5 1/2 inches
Wing area: 608 square feet.

Weights: 27,000 pounds empty(F-15A) 28,200 pounds empty (F-15C), 40,000 pounds combat, 41,500 pounds gross, 66,000 pounds maximum takeoff (F-15A) 68,470 pounds maximum takeoff (F-15C).

Fuel: Maximum internal fuel 1790 US gallons. Three 610-gallon drop tanks can be carried, one on the fuselage centerline and one on each of the underwing pylons, bringing total fuel capacity to 3620 US gallons. The F-15C and D can carry Conformal Fuel Tanks (CFTs). Each CFT carries an additional 849 US gallons of fuel.

Armament: One 20-mm General Electric M61A1 Vulcan cannon n the right wing root with 940 rounds. Provision for four AIM-7F/M Sparrow or four AIM-120 AMRAAM missiles on hardpoints attached to the lower outer edges of the air intake trunks (or the CFTs C and D only), two on each side. Four AIM-9 Sidewinders infrared-homing missiles are carried on the underwing pylons, two on each side.

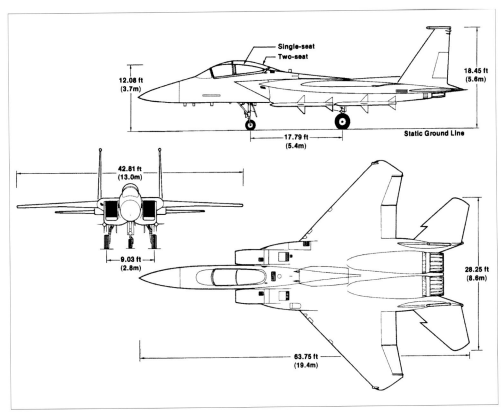

Specification of McDonnell Douglas F-15E Strike Eagle

Two Pratt & Whitney F100-PW-229 turbofans, each rated at 17,800 lb.s.t. dry and 29,100 lb.s.t. with afterburning.

Performance: Maximum speed Mach 2.54 (1676 mph) at 40,000 feet (short-endurance dash), Mach 2.3 (1520 mph) (sustained). Maximum combat radius 790 miles. Maximum ferry range 2765 miles.

Weights: 31,700 pounds empty, 84,000 pounds maximum takeoff.

Dimensions: Wingspan 42 feet 9 3/4 inches
Length: 63 feet 9 inches
Height: 18 feet 5 1/2 inches
Wing area: 608 square feet.

Armament: One 20-mm M61A1 rotary cannon with 450 rounds. A maximum ordnance load of 24,500 pounds can be carried on the centerline and two underwing stations plus four tangential carriers attached to the conformal fuel tanks. In the air-to-air mission, up to four AIM-9L/M Sidewinders can be carried on the underwing stations and four AIM-7F/M Sparrow missiles can be carried on the conformal fuel tank attachments. Alternatively, up to eight AIM-120 AMRAAM missiles can be carried.

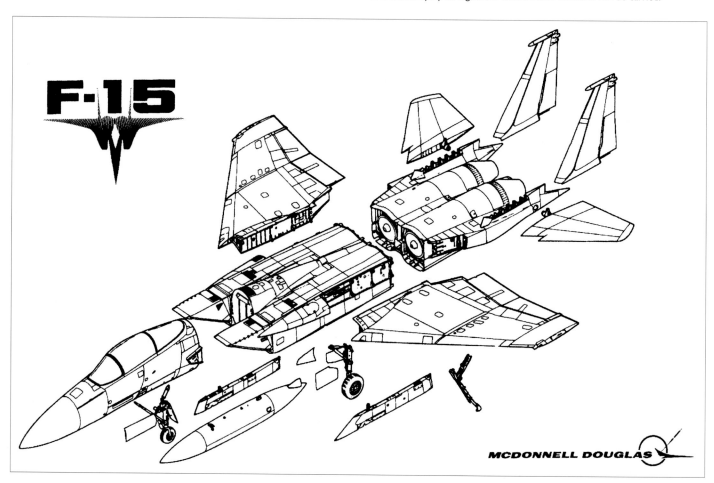

USAF F-15 SERIAL NUMBERS

Though F-15s are operated by the Israeli Air Force, the Japanese Self Defense Force, and the Royal Saudi Air Force, only the serial numbers of the USAF F-15 are listed in this table.

F-15As

71-0280/0281	McDonnell Douglas F-15A-1-MC	0281 bailed to NASA in 1975. Returned to USAF in 1983, now on display at Langley AFB
71-0282/0284	McDonnell Douglas F-15A-2-MC	0284 to GF-15A
71-0285/0286	McDonnell Douglas F-15A-3-MC	0286 to GF-15A
71-0287/0289	McDonnell Douglas F-15A-4-MC	0287 bailed to NASA in 1976 as 835
72-0113/0116	McDonnell Douglas F-15A-5-MC	0116 delivered to Israel, Peace Fox I
72-0117/0120	McDonnell Douglas F-15A-6-MC	0117, 0118 delivered to Israel, Peace Fox I 0119 Streak Eagle. This plane is on display at WPAFB Museum. 120 was delivered to Israel, Peace Fox I
73-0085/0089	McDonnell Douglas F-15A-7-MC	
73-0090/0097	McDonnell Douglas F-15A-8-MC	
73-0098/0107	McDonnell Douglas F-15A-9-MC	
74-0081/0093	McDonnell Douglas F-15A-10-MC	
74-0094/0111	McDonnell Douglas F-15A-11-MC	
74-0112/0136	McDonnell Douglas F-15A-12-MC	
74-0143/0157	McDonnell Douglas F-15A/B	Cancelled contract
75-0018/0048	McDonnell Douglas F-15A-13-MC	
75-0049/0079	McDonnell Douglas F-15A-14-MC	
75-0090/0124	McDonnell Douglas F-15A/B	Cancelled contract
76-0008/0046	McDonnell Douglas F-15A-15-MC	
76-0047/0083	McDonnell Douglas F-15A-16-MC	
76-0084/0113	McDonnell Douglas F-15A-17-MC	0086 used for trials with Vought ASM-135A ASAT.
76-0114/0120	McDonnell Douglas F-15A-18-MC	120 was delivered to Israel
76-0121/0123	McDonnell Douglas F-15A	Canceled contract
77-0061/0084	McDonnell Douglas F-15A-18-MC	0084 used as test bed for APG-63 radar
77-0085/0119	McDonnell Douglas F-15A-19-MC	
77-0120/0153	McDonnell Douglas F-15A-20-MC	

TF-15As/F-15Bs

Serial	Designation	Notes
71-0290	McDonnell Douglas F-15B-3-MC	Later modified as part of STOL and Maneuver Technology Demonstrator program (Agile Eagle)
71-0291	McDonnell Douglas F-15B-4-MC	Used for evaluation of FAST Pack conformal fuel tanks and LANTIRN pod. Also became development aircraft for F-15E Strike Eagle
73-0108/0110	McDonnell Douglas F-15B-7-MC	
73-0111/0112	McDonnell Douglas F-15B-8-MC	
73-0113/0114	McDonnell Douglas F-15B-9-MC	
74-0137/0138	McDonnell Douglas F-15B-10-MC	
74-0139/0140	McDonnell Douglas F-15B-11-MC	
74-0141/0142	McDonnell Douglas F-15B-12-MC	74-0141 was operated by NASA as NASA 836
74-0143/0157	McDonnell Douglas F-15A/B	Canceled contract
75-0080/0084	McDonnell Douglas F-15B-13-MC	
75-0085/0089	McDonnell Douglas F-15B-14-MC	
75-0090/0124	McDonnell Douglas F-15A/B	Canceled contract
76-0124/0129	McDonnell Douglas F-15B-15-MC	
76-0130/0135	McDonnell Douglas F-15B-16-MC	
76-0136/0140	McDonnell Douglas F-15B-17-MC	
76-0141/0142	McDonnell Douglas F-15B-18-MC	
77-0154/0156	McDonnell Douglas F-15B-18-MC	
77-0157/0162	McDonnell Douglas F-15B-19-MC	
77-0163/0168	McDonnell Douglas F-15B-20-MC	0166 used as test vehicle for Integrated Flight Control/Firefly III program

F-15Cs

Serial	Designation	Notes
78-0468/0495	McDonnell Douglas F-15C-21-MC	
78-0496/0522	McDonnell Douglas F-15C-22-MC	
78-0523/0550	McDonnell Douglas F-15C-23-MC	
78-0551/0560	McDonnell Douglas F-15C	Canceled contract
79-0015/0037	McDonnell Douglas F-15C-24-MC	0015, 0017/0019, 0023, 0024, 0028, 0031/0033 transferred to Saudi Arabia
79-0038/0058	McDonnell Douglas F-15C-25-MC	0038, 0039, 0043, 0045, 0051, 0052, 0055 transferred to Saudi Arabia
79-0059/0081	McDonnell Douglas F-15C-26-MC	0060, 0062, 0063 transferred to Saudi Arabia
80-0002/0023	McDonnell Douglas F-15C-27-MC	
80-0024/0038	McDonnell Douglas F-15C-28-MC	
80-0039/0053	McDonnell Douglas F-15C-29-MC	
81-0020/0031	McDonnell Douglas F-15C-30-MC	
81-0032/0040	McDonnell Douglas F-15C-31-MC	
81-0041/0056	McDonnell Douglas F-15C-32-MC	
81-0057/0060	McDonnell Douglas F-15C	Canceled contract
82-0008/0022	McDonnell Douglas F-15C-33-MC	
82-0023/0038	McDonnell Douglas F-15C-34-MC	

83-0010/0034	McDonnell Douglas F-15C-35-MC	
83-0035/0043	McDonnell Douglas F-15C-36-MC	
83-0044/0045	McDonnell Douglas F-15C	Canceled contract
84-0001/0015	McDonnell Douglas F-15C-37-MC	
84-0016/0031	McDonnell Douglas F-15C-38-MC	
84-0032/0041	McDonnell Douglas F-15C	Canceled contract
85-0093/0107	McDonnell Douglas F-15C-39-MC	
85-0108/0128	McDonnell Douglas F-15C-40-MC	
86-0143/0162	McDonnell Douglas F-15C-41-MC	
86-0163/0180	McDonnell Douglas F-15C-42-MC	

F-15Ds

78-0561/0565	McDonnell Douglas F-15D-21-MC	
78-0566/0570	McDonnell Douglas F-15D-22-MC	
78-0571/0574	McDonnell Douglas F-15D-23-MC	
78-0575	McDonnell Douglas F-15D	Canceled contract
79-0004/0006	McDonnell Douglas F-15D-24-MC	All transferred to Saudi Arabia
79-0007/0011	McDonnell Douglas F-15D-25-MC	
79-0012/0014	McDonnell Douglas F-15D-26-MC	
80-0054/0055	McDonnell Douglas F-15D-27-MC	
80-0056/0057	McDonnell Douglas F-15D-28-MC	
80-0058/0061	McDonnell Douglas F-15D-29-MC	
81-0061/0062	McDonnell Douglas F-15D-30-MC	
81-0063/0065	McDonnell Douglas F-15D-31-MC	
81-0066/0067	McDonnell Douglas F-15D	Canceled contract
82-0044/0045	McDonnell Douglas F-15D-33-MC	
82-0046/0048	McDonnell Douglas F-15D-34-MC	
83-0046/0048	McDonnell Douglas F-15D-35-MC	
83-0049/0050	McDonnell Douglas F-15D-36-MC	
84-0042/0044	McDonnell Douglas F-15D-37-MC	
84-0045/0046	McDonnell Douglas F-15D-38-MC	
84-0047/0048	McDonnell Douglas F-15D	Canceled contract
85-0129/0134	McDonnell Douglas F-15D-39-MC	
86-0181/0182	McDonnell Douglas F-15D-41-MC	

F-15Es

71-0291	McDonnell Douglas F-15B-4-MC	This F-15B was used for evaluation of FAST Pack conformal fuel tanks and LANTIRN pod. Also became development aircraft for F-15E Strike Eagle
86-0183/0184	McDonnell Douglas F-15E-41-MC	
86-0185/0190	McDonnell Douglas F-15E-42-MC	
87-0169/0189	McDonnell Douglas F-15E-43-MC	
87-0190/0210	McDonnell Douglas F-15E-44-MC	
87-0211/0216	McDonnell Douglas F-15E-44-MC	Canceled contract
88-1667/1687	McDonnell Douglas F-15E-45-MC	
88-1688/1708	McDonnell Douglas F-15E-46-MC	
89-0046/0063	McDonnell Douglas F-15E-47-MC	Canceled contract
89-0064/0081	McDonnell Douglas F-15E-48-MC	Canceled contract
89-0471/0488	McDonnell Douglas F-15E-47-MC	
89-0489/0506	McDonnell Douglas F-15E-48-MC	
90-0227/0244	McDonnell Douglas F-15E-49-MC	
90-0245/0262	McDonnell Douglas F-15E-50-MC	
91-0300/0317	McDonnell Douglas F-15E-51-MC	
91-0318/0335	McDonnell Douglas F-15E-52-MC	
91-0600/0605	McDonnell Douglas F-15E-53-MC	
92-0364/0366	McDonnell Douglas F-15E-53-MC	
96-0200/0205	McDonnell Douglas F-15E-58-MC	
97-0217/0222	McDonnell Douglas F-15E-61-MC	
98-0131/0135	McDonnell Douglas F-15E-62-MC	

USAF F-15 LOSSES

1	14 Oct 75	73-0088	A	LA	555 TFTS	Crashed at Luke AFB, Arizona.
2	28 Feb 77	74-0129	A	WA	57 FWW	Crashed on range following collision with F-5E
3	06 Dec 77	75-0085	B	WA	57 FWW	Crashed on Nellis AFB Ranges, Nevada.
4	08 Feb 78	73-0097	A	LA	555 TFTS	Written off due to ground incident.
5	17 Apr 78	75-0059	A	BT	525 TFS	Crashed into North Sea off the coast of Cromer, Norfolk, United Kingdom.
6	15 Jun 78	76-0047	A	BT	53 TFS	Crashed into North Sea during ACT.
7	06 Jul 78	76-0053	A	BT	53 TFS	Crashed near Daun, Germany.
8	01 Sep 78	75-0018	A	FF	71 TFS	Crashed into the Atlantic, Virginia Capes
9	19 Dec 78	75-0063	A	BT	525 TFS	Crashed near Ahlhorn, Schleswig Holstein, Germany.
10	28 Dec 78	75-0064	A	BT	36 TFW	Crashed near Daun, Germany.
11	29 Dec 78	74-0136	A	WA	57 FWW	Crashed at Nellis AFB, Nevada.
12	16 Feb 79	77-0107	A	HO	9 TFS	Crashed on Nellis AFB Ranges, Nevada.
13	13 Mar 79	77-0076	A	HO	9 TFS	
14	25 Apr 79	77-0167	B		MDC	Crashed near Fredericktown, Missouri, on test flight.
15	03 Jun 79	76-0035	A	BT	36 TFW	Crashed on landing at Bitburg AB, Germany.
16	14 Sep 79	76-0085	A	WA	57 FWW	
17	03 Oct 79	77-0072	A	HO	49 TFW	Mid-air collision with F-15A 77-061 HO, NAS Fallon, Nevada ranges
18	04 Mar 80	75-0070	A	BT	36 TFW	Crashed near Baden-Baden, Germany.
19	06 Mar 80	76-0082	A	BT	36 TFW	Crashed near Bitburg, Germany.
20	10 Mar 80	75-0023	A	FF	27 TFS	Destroyed by fire on flight line at Langley AFB, Virginia.
21	25 Jul 80	76-0013	A	BT	36 TFW	
22	21 Jan 81	77-0164	B	WA	57 FWW	Mid-air collision with F-5E 74-1517 57 FWW over Nellis AFB, Nevada ranges.
23	17 Feb 81	76-0065	A	LA	405 TTW	Crashed into Pacific Ocean.
24	23 Jun 81	79-0040	C	BT	36 TFW	Crashed 15 miles from Bremen, Germany.
25	12 Sep 81	80-0007	A	BT	22 TFS	Crashed at Soesterberg, The Netherlands during aerial demonstration.
26	02 Nov 81	75-0051	A	EG	33 TFW	Near Panama City, Florida; mid-air collision with an F-15 on a refueling mission.
27	15 Dec 81	73-0106	A	LA	461 TFTS	Nellis AFB, Nevada Range Complex tc \II "Nellis AFB Range Complex
28	06 Apr 82	78-0524	C	ZZ	18 TFW	Crashed into Pacific Ocean 40 miles NW of Okinawa
29	22 Dec 82	80-0025	C	BT	53 TFS	Crashed near Herschbach, Germany.
30	29 Dec 82	78-0481	C	ZZ	18 TFW	Crashed into Pacific Ocean, 92 miles NE of Okinawa, Japan, after colliding with F-15C 78-0540
31	29 Dec 82	78-0540	C	ZZ	18 TFW	Crashed into Pacific Ocean, 92 miles NE of Okinawa, Japan, after colliding with F-15C 78-0481
32	04 Jan 83	80-0036	C	FF	1 TFW	
33	04 Feb 83	76-0081	A	EG	33 TFW	
34	09 May 83	77-0094	A	HO	49 TFW	Crashed at White Sands missile range, New Mexico.
35	01 Jun 83	79-0071	C	BT	36 TFW	Crashed near Kusel, Germany, after mid-air collision with F-15C 80-008 BT.
36	01 Jun 83	80-0008	C	BT	36 TFW	Crashed near Kusel, Germany, after mid-air collision with F-15C 79-071
37	06 Oct 83	75-0076	A	EG	33 TFW	Crashed 45 miles N of Cold Lake, Canada, after colliding with F-5E 74-1509 57 FWW
38	09 Mar 84	74-0094	A	AK	21 TFW	Crashed in Alaska.
39	10 Apr 84	79-0044	C	BT	36 TFW	Crashed near Lommersdorf, Rheinland-Pfaltz, Germany.
40	17 Aug 84	74-0139	B	AK	21 TFW	
41	21 Aug 84	75-0087	B	TY	325 TTW	Mid-air collision with F-4E 68-0535 RS, 526 TFS
42	20 Mar 85	74-0120	A	AK	43 TFS	
43	24 Jun 85	74-0087	A	AK	43 TFS	Crashed into Yukon river, Alaska.
44	05 Nov 85	74-0090	A	AK	43 TFS	
45	16 Dec 85	84-0042	D	AD	3246 TW	Crashed into Gulf of Mexico.
46	02 Jan 86	80-0037	C	IS	57 FIS	Crashed into Atlantic Ocean.
47	07 Jan 86	79-0061	C	BT	525 TFS	Crashed near Rimschweiler, Germany after mid-air collision with F-15C 80-0032 525 TFS "BT".
48	07 Jan 86	80-0032	C	BT	525 TFS	Crashed near Rimschweiler, Germany after mid-air collision with F-15C 79-0061 36 TFW "BT".
49	14 Jan 86	76-0023	A	–	5 FIS	Crashed in the Guadalupe Mountains, New Mexico.
50	07 Mar 86	76-0055	A	LA	405 TTW	Mid-air collision with F-15A 76-074 LA
51	07 Mar 86	76-0074	A	LA	405 TTW	Mid-air collision with F-15A 76-055 LA
52	09 Jun 86	78-0472	C	ZZ	18 TFW	
53	12 Sep 86	77-0153	A	HO	49 TFW	Mid-air with another F-15 that landed safely.
54	09 Mar 87	77-0075	A	HO	49 TFW	Crashed 3 miles SE of Holloman AFB, New Mexico
55	19 May 87	78-0495	C	ZZ	18 TFW	Crashed into Pacific Ocean.
56	08 Jun 87	81-0056	C	FF	1 TFW	Crashed in Virginia.
57	02 Oct 87	75-0027	A	TY	325 TTW	Crashed in Apalachicola Forest, Alabama/Georgia.
58	24 Nov 87	75-0056	A		128 TFS	Collided with F-16B 79-0419 466 TFS "HI" near Wadley, Georgia.
59	08 Nov 88	80-0017	C	AK	21 TFW	Crashed 5 miles NW of Kodiak, Alaska.
60	01 May 89	76-0138	B	TY	95 TFTS	Crashed into Gulf of Mexico 65 miles SE of Tyndall AFB, Florida.
61	06 Jul 89	85-0109	A	EG	33 TFW	
62	10 Aug 89	77-0101	A	HO	49 TFW	Crashed 60 miles N of Holloman AFB, White Sands missile range, New Mexico.
63	06 Nov 89	84-0029	A	WA	57 FWW	Crashed 60 miles N of Las Vegas, Nevada.
64	28 Dec 89	86-0153	C	EG	33 TFW	Crashed into Gulf of Mexico, 40 miles SE of Apalachicola, Florida.
65	16 Jan 90	80-0059	D			

66	24 Jan 90	78-0534	C	ZZ	18 TFW	Mid-air collision with F-15C 78-520 ZZ, crashed into ocean 50 miles NW of Clark AB, Philippines. 78-0520 returned to base and was repaired
67	15 Mar 90	76-0069	A	LA	426 TFTS	Crashed 70 miles N of Phoenix, Arizona.
68	25 Apr 90	81-0049	C	CR	32 TFS	Crashed 9 miles off the coast of England into the North Sea.
69	30 Sep 90	87-0203	E	SJ	4 TFW	Crashed during operation Desert Shield
70	24 Oct 90	79-0067	C	BT	22 TFS	Crashed into Mediterranean 30 miles from Decimomannu, Italy.
71	18 Jan 91	88-1689	E	SJ	336 TFS	COMBAT (Desert Storm)
72	20 Jan 91	88-1692	E	SJ	335 TFS	COMBAT (Desert Storm)
73	27 Mar 91	78-0526	C	ZZ	18 TFW	
74	16 Sep 91	87-0172	E	LA	405 TTW	
75	15 Jan 92	75-0071	A		128 TFS	Mid-air collision with F-15A 75-0075 128 TFS. 75-0075 was damaged, repaired and returned to service.
76	21 Jan 92	81-0052	C	WA	57 FWW	Crashed somewhere in the Nevada desert.
77	22 Apr 92	80-0023	C	BT	22 FS	Crashed near Stuttgart, Germany
78	13 Jul 92	85-0116	C	EG	60 FS	Crashed into the Gulf of Mexico, 90 miles S of Eglin AFB, Florida.
79	10 Aug 92	89-0479	E	WA	57 FWW	Crashed NE of Las Vegas, Nevada
80	30 Nov 92	83-0021	C	FF	71 FS	Crashed into Persian Gulf near Dhahran, Saudi Arabia.
81	15 Mar 93	79-0027	C	TY	95 FS	Crashed into Gulf of Mexico.
82	12 Jun 93	77-0117	A		122 FS	Crashed 32 miles E of NAS New Orleans, Louisiana.
83	17 Dec 93	75-0054	A	GA	128 FS	Mid-air collision with F-16A 184 FS
84	04 Apr 94	78-0497	C	ZZ	44 FS	Crashed shortly after takeoff from Kadena AB
85	05 May 94	79-0058	C	TY	1 FS	G-induced loss of consciousness (G-LOC)
86	06 May 94	78-0530	C	ZZ	67 FS	Crashed after collision with F-16C 87-274 WP, 80 FS, 2 miles off the coast of South Korea.
87	18 Apr 95	89-0504	E	SJ	336 FS	Crashed into sea off the coast of North Carolina during air-to-air training with four other F-15Es. WSO killed, pilot OK.
88	30 May 95	79-0068	C	SP	53 FS	Crashed on takeoff at Spangdahlem AFB, Germany. Pilot Major Donald G. Lowry died during transportation to the local hospital, cause of the crash was a maintenance error – the ailerons and elevator connections were incorrectly installed.
89	03 Aug 95	78-0537	C	ZZ	67 FS	Crashed 100 miles E of Elmendorf AFB, Alaska in Yukon-Charley Rivers National Preserve, Alaska during "Cope Thunder" from Elmendorf AFB, Alaska, Pilot Capt. Garth Doty ejected and was rescued, 'Duster 3'.
90	18 Oct 95	78-0529	C	ZZ	44 FS	Crashed 60 miles S of Okinawa, Japan; pilot Capt. Donald McKercher was rescued by a JASDF helicopter from Naha.
91	21 Mar 96	82-0023	C	FF	27 FS	Crashed on takeoff from Nellis AFB, Nevada during 'Green Flag'.
92	01 May 96	76-0061	A	SL	110 FS	Engine fire, landed 100 knots too fast at Whiteman AFB, Missouri, and ran off the end of the runway.
93	27 Aug 96	86-0150	C	MO	390 FS	Crashed during a routine mission from Mountain Home AFB, Idaho the pilot 1st Lt. Evan Dertien ejected successfully. (Callsign: 'Bacon 04')
94	10 Jan 97	85-0099	C	EG	58 FS	
95	11 Jul 97	89-0491	E	SJ	334 FS	Engine fire, near Dare County range, North Carolina
96	24 Nov 97	83-0033	C	FF	94 FS	50 miles off the coast of Virginia as a result of technical problems 1LT David M. Nyikos ejected, rescued by Coast Guard helicopter an hour later.
97	08 Jun 98	77-0120	A	JZ	159 FW	Crashed at JRB New Orleans, Louisiana.
98	20 Oct 98	89-0497	E	MO	366 WG	Crashed near McDermott, Oregon during a night training mission. Both crew members were killed.
99	28 Jan 99	80-0011	C	OT	85 FLTS	Mid air with 82-0022 over the Gulf of Mexico. Pilot ejected successfully.
100	28 Jan 99	82-0022	C	OT	85 FLTS	Mid air with 80-0011 over the Gulf of Mexico. Pilot ejected successfully.
101	15 Jun 99	79-0013	D	ED	445 FTLS	Mid air with 84-0008 60 miles east of Tonopah, Nevada before starting test phase of a computer Y2K test mission. Pilot ejected safely.
102	15 Jun 99	82-0008	C	OT	422 FLTS	Mid air with 79-0013 60 miles east of Tonopah, Nevada before starting test phase of a computer Y2K test mission. Pilot ejected safely.
103	19 Aug 99	76-0117	A	SL	131 FW	Aircraft was lost on an air-to-air practice mission. The pilot ejected safely.